Einführung in die Quantenmechanik und ihre Anwendungen

Von

Prof. Dr. P. Gombás

und

Dr. D. Kisdi

Physikalisches Institut der Universität für Technische Wissenschaften, Budapest

36 Abbildungen. 251 Seiten. 1970.

Ganzleinen

S 308,—, DM 49,—, US $ 12.25

Dieses Buch gibt eine Einführung in die Quantenmechanik, insbesondere in die Wellenmechanik, mit zahlreichen Anwendungen. Es gliedert sich in zwei Teile. Im ersten Teil, der die Grundlagen behandelt, werden nach den experimentellen grundlegenden Resultaten und einem kurzen Überblick über die Bohrsche Quantentheorie die Grundlagen der Quantenmechanik gebracht, und zwar hauptsächlich das, was im zweiten Teil zu den Anwendungen gebraucht wird. Der zweite Teil enthält zahlreiche Anwendungen, und zwar zunächst auf Einteilchenprobleme, danach auf Streuprobleme und schließlich Anwendungen von Näherungsmethoden auf einfache Probleme. Die Darstellungsweise des Stoffes ist möglichst einfach und bezweckt, den Leser mit geringer Mühe in die Rechenmethoden der Wellenmechanik einzuführen und mit der Durchrechnung der meistens bis ins kleinste Detail ausgearbeiteten Anwendungen zum Selbststudium anzuregen. Das Buch kann somit auch als Lehrbuch zur Einführung in die Wellenmechanik dienen.

Topics in Applied Quantumelectrodynamics

Paul Urban

1970

Springer-Verlag

Wien · New York

Professor Dr. Paul Urban
Institute of Theoretical Physics
University of Graz

With 55 Figures

ISBN-13:978-3-7091-8249-9 e-ISBN-13:978-3-7091-8247-5
DOI: 10.1007/978-3-7091-8247-5

Softcover reprint of the hardcover 1st edition 1970
Library of Congress Catalog Card Number 71-98304

Title No. 9260

Preface

These lectures represent a condensation of a number of colloquia, seminars and discussions held at the Institute of Theoretical Physics of the University of Graz during the last years and epitomize the principal lines of research undertaken by my group. From the very beginning of my appointment at the University of Graz in 1947 I have been concerned with the task of bringing up a relatively small group of scientifically interested and open-minded co-workers and of stimulating them to sound scientific research. Since 1930 I myself have dealt with subjects of the kind treated in these lectures, to which I was introduced by my late friend and teacher Th. Sexl. But also as assistant and co-worker of E. Fues and H. Thirring I frequently worked on these problems, constantly using new methods and lines of approach. During the last years of the war and the first ones afterwards I had the fortunate opportunity to receive many stimulating ideas and comments on my work from A. Sommerfeld on the occasion of my frequent visits to Munich. Especially this last period, although partially connected with personal difficulties and troubles of many kinds stemming from the turbulence of lost-war readjustments, I consider to be one of the most valuable times in my life. The experiences which I accumulated then I later tried to put into effect, at least to a limited extent, in order to create a productive climate for research in the spirit of A. Sommerfeld and his school of thought. The number of my students who already hold respected positions in the scientific community at least give me the confidence that my work and my efforts were not in vain.

The following lectures are divided into two parts:

1. Electron Scattering and Nucleon Form Factors
2. Radiative Corrections

It was my intention in writing this summary not only to refer to the work done at my Institute but also to give an account of related research of many colleagues which seemed important to me. In addition, experimental results are frequently included for comparison together with a discussion of the deviations which occasionally appear. Especially the second part contains basic computations which are required for the design of experimental arrangements of current interest.

In the compilation of the text I was assisted in manifold ways by the members of my Institute. The scientific achievements of these co-workers are documented by their papers included in the reference list and give evidence of their diligence and talent. In editing these lectures especially Mr. P. PESEC and Mr. F. WIDDER were of dedicated help. I would also like to thank my friends and colleagues Prof. T. ERBER and Prof. R. ROHRLICH for a critical reading of the manuscript. The typing of manuscript was done perfectly and within shortest time by my secretary of many years, Mrs. ANNELIESE KÜHNELT; with the same skill she is doing so for our annual "Schladminger Universitätswochen". To all of them I want to express my sincerest thanks.

Graz, Fall 1969 PAUL URBAN

Contents

Part I

Electron-Scattering and Nucleon Form Factors

I. The Dirac-Foldy-Wouthuysen Transformation 1
1. The Dirac Equation 1
2. The DFW Transformation (Free Case) 6
3. The Foldy-Heisenberg (F-H) Picture 10
4. The Foldy-Interaction (F-I) Picture 14
5. Connection between DFW and Lorentz Transformation 18
6. Connection between F-H and F-I Picture 20
7. Ambiguities of the F-H Picture 21
8. Application of the DFW Transformation 22
Literature 24

II. Determination of Proton-Form Factors Derived from Electron-Proton-Scattering 25
1. Introductory Remarks 25
2. Relativistic Electron-Proton-Scattering (Rosenbluth Formula) 27
3. Electromagnetic Form Factors of the Proton 38
Literature 44

III. Determination of the Neutron Form Factors Derived from Quasielastic Electron-Deuteron-Scattering 45
1. Introduction 45
2. Quasielastic Electron-Deuteron-Scattering 48
3. Calculation of the Cross-Section in the Lab.-System 52
4. Influence of the Deuteron Model 67
5. Final State Corrections 71
Appendix 77
Literature 79

IV. Calculation of Nucleon Form Factors in Dispersion Theory 81
Literature 95

Part II

Radiative Corrections

Introduction 96

I. Classical Radiation of Long-Wavelength Photons 101
1. Introduction 101
2. Emission of Soft Photons 106
3. Experimental Cross-Section and Radiative Corrections 116
4. Calculations of Radiative Corrections 126

II. Summation over Soft-Photon Contributions 145
1. Separation of Soft and Hard Photons 145
2. Virtual Soft Photons 152
3. Real Soft Photons 156
4. Summation over Soft Photons 160

III. Radiative Corrections in the Framework of Quantumelectrodynamics 168
1. Infrared Divergences 169
2. The Canceling of Infrared Divergences 176
3. Summation of the "Infrared" Contributions 179

IV. Examples 186
1. Electron-Proton-Scattering 186
2. Inelastic Electron-Scattering 216
Literature 226

Appendix

A. The Green-Functions of the Klein-Gordon Equation and the Dirac Equation 228

B. Theory of Bosons, Klein-Gordon Equation 234
1. Derivation of the Klein-Gordon Equation 234
2. Scalar Wave Functions 235
3. Wave Functions for Particles with Spin 1 240

C. Theory of Fermions, Dirac Equation 247
1. Relativistic Wave-Equation for Fermions 247
2. Charge Conjugation 250
3. Solutions for a Free Particle 253

Part I
ELECTRON-SCATTERING AND NUCLEON FORM FACTORS

I. The Dirac-Foldy-Wouthuysen Transformation+

1. The Dirac Equation

The Dirac-Foldy-Wouthuysen (DFW) transformation will be discussed in its application to the Dirac equation, where it is best known; essentially the same conclusions, however, also hold in case of the Klein-Gordon equation [2], [3].

The relativistic motion of spin $\frac{1}{2}$ - particles in an electromagnetic potential $A_\mu(x)$ is governed by the Dirac equation; with the notation ($\hbar = c = 1$)

$$(ab) = a^\mu b_\mu = a_\mu b^\mu = a^\mu b^\nu g_{\mu\nu} = a^0 b^0 - \vec{a}\cdot\vec{b} \; ;$$

$$\gamma^\mu = (\gamma^0, \vec{\gamma}) \; ;$$

$$\not{a} = a^\mu \gamma_\mu \; ;$$

$$\partial_\mu = \frac{\partial}{\partial x^\mu} \; ;$$

$$\gamma^0 = \beta = \begin{pmatrix} I & 0 \\ 0 & -I \end{pmatrix}, \qquad \vec{\gamma} = \beta\vec{\alpha}, \qquad \vec{\alpha} = \begin{pmatrix} 0 & \vec{\sigma} \\ \vec{\sigma} & 0 \end{pmatrix} \; ; \tag{1,1}$$

+ This transformation has been discussed by P.A.M. Dirac [1] as early as in 1934 (private communication by Prof. P.E. Wigner).

$$I = \begin{pmatrix} 1 & 0 \\ 0 & 1 \end{pmatrix}, \quad \sigma_1 = \begin{pmatrix} 0 & 1 \\ 1 & 0 \end{pmatrix}, \quad \sigma_2 = \begin{pmatrix} 0 & -i \\ i & 0 \end{pmatrix}, \quad \sigma_3 = \begin{pmatrix} 1 & 0 \\ 0 & -1 \end{pmatrix};$$

it has the form $[A_\mu = (\Phi, \vec{A})]$:

$$(i\not\partial - e\not{A}(x) - m)\,\psi(x) = 0\,. \tag{1,2}$$

Here $\psi(x)$ is a four-component spinor, γ^μ we can take as in the definition (1,1) or according to an equivalent representation; in any case they fulfil the anticommutation relation

$$\{\gamma^\mu, \gamma^\nu\} \equiv \gamma^\mu \gamma^\nu + \gamma^\nu \gamma^\mu = 2g^{\mu\nu}\,. \tag{1,3}$$

By the iteration of (1,2) we obtain a wave equation which differs from the Klein-Gordon equation by the coupling between spin and electromagnetic field:

$$(i\not\partial - e\not{A}(x) + m)(i\not\partial - e\not{A}(x) - m)\,\psi(x) =$$

$$= [(i\partial_\mu - eA_\mu)(i\partial^\mu - eA^\mu) - m^2 + \frac{e}{2} F^{\mu\nu}\sigma_{\mu\nu}]\,\psi(x) = 0\,, \tag{1,4}$$

where we used

$$F_{\mu\nu} = \partial_\nu A_\mu - \partial_\mu A_\nu, \quad F^{ok} = E^k, \quad F^{kl} = \epsilon_{klm} B^m\,;$$

$$\sigma_{\mu\nu} = \frac{i}{2}[\gamma_\mu, \gamma_\nu] = \frac{i}{2}(\gamma_\mu\gamma_\nu - \gamma_\nu\gamma_\mu)\,.$$

The term $\frac{e}{2} F^{\mu\nu}\sigma_{\mu\nu}$ in (1,4) corresponds to the term $\frac{e}{m}\vec{\sigma}\cdot\vec{B}$ of the nonrelativistic Pauli equation. In momentum space, through the transformation

$$\psi(x) = \frac{1}{(2\pi)^2}\int d^4p\, e^{-ipx}\,\psi(p)\,, \tag{1,5}$$

equation (1,2) has the form

$$(\not{p} - e\not{A} - m)\psi(p) = 0\,, \tag{1,6}$$

which for free particles reads

$$(\not{p} - m)\,\psi(p) = 0\,. \tag{1,7}$$

Equation (1,7) is a short-hand notation for actually four homogeneous linear equations involving the four components of the spinor $\psi(p)$; the determinant of coefficients is

$$\det (\not{p} - m) = (p^2 - m^2)^2 = (p_0 - E)^2 (p_0 + E)^2 , \qquad (1,8)$$

where $E = +\sqrt{\vec{p}^2 + m^2}$. Equation (1,7) therefore possesses two solutions to each value of p_0 ($\pm E$), corresponding to the spin orientations.

As usual the solutions corresponding to negative energy ($p_0' = -E$) are interpreted as antiparticles in the framework of Dirac's hole-theory which establishes a complete symmetry between particles and antiparticles (charge-conjugation symmetry).

Exactly this existence of antiparticles, which is necessarily connected with each local relativistic covariant wave equation (this is part of the conclusions drawn from the PCT theorem), leads to difficulties in the interpretation and application of the Dirac equation.

Writing (1,7) in Hamiltonian form

$$P_0 \psi = H \psi = (\vec{\alpha} \cdot \vec{p} + \beta m) \psi , \qquad (1,9)$$

(in the following p_μ may stand for either p_μ or $i\partial_\mu$) we can for example investigate the operator $\dot{\vec{x}}$ which may be interpreted as velocity operator:

$$\vec{v} = \dot{\vec{x}} = i[H, \vec{x}] = \vec{\alpha} , \qquad (1,10)$$

by virtue of

$$[x_k , p_l] = i\delta_{kl} . \qquad (1,11)$$

Since $\alpha_i^2 = 1$ the magnitude of the velocity $\vec{v}$ is always equal to the velocity of light c . In addition, the different components of the velocity cannot be defined simultaneously since $[\alpha_i , \alpha_j] \neq 0$. This, however, contradicts the possibility of observation of the velocity. Further problems arise in applications if one wants to employ nonrelativistic wave functions, e.g. from nuclear physics, together with the Dirac equation.

Thus it is evident that another representation for the Dirac equation has to be found in order to make possible a physical interpretation. A Dirac-particle with positive energy has to be represented by only two vectors in Hilbert space corresponding to its two possible spin orientations. Therefore two components of the four-component wave function in Dirac's theory are superfluous, and we have to find a transformation reducing the Dirac equation to a two-component equation, for example the nonrelativistic Pauli-theory.

Large and Small Components, Pauli Equation

The Dirac equation (1,6) can be rewritten as two coupled equations by expressing the wave function in terms of two-component spinors φ and χ:

$$\psi = \begin{bmatrix} \varphi \\ \chi \end{bmatrix} ; \tag{1,12}$$

$$(\vec{\sigma} . (\vec{p} - e\vec{A})) \chi + (e\Phi + m) \varphi = E\varphi,$$

$$(\vec{\sigma} . (\vec{p} - e\vec{A})) \varphi + (e\Phi - m) \chi = E\chi . \tag{1,13}$$

Solving the last equation for χ we get

$$\chi = \frac{1}{2[m + \frac{1}{2}(E - e\Phi - m)]} \vec{\sigma} . (\vec{p} - e\vec{A}) \varphi , \tag{1,14}$$

and inserting this in (1,13) we arrive at an equation exactly equivalent to the Dirac equation:

$$[\vec{\sigma}(\vec{p} - e\vec{A}) \frac{1}{2[m + \frac{1}{2}(E - e\Phi - m)]} \vec{\sigma} . (\vec{p} - e\vec{A}) + e\Phi] \varphi = (E-m) \varphi . \tag{1,15}$$

In the nonrelativistic limit we have

$$E - m,\ e\Phi,\ \vec{p},\ e\vec{A} \ll m\ ;\ m + \tfrac{1}{2}(E - e\Phi - m) \approx m,$$

and therefore also

$$\chi \ll \varphi,$$

or, in other words, the ratio of χ and φ is of order $\frac{p}{m}$ or $\frac{v}{c}$. Hence χ and φ are called small and large components respectively. The terms "even" and "odd" operator are closely connected with this:

An operator is called "even" if it contains no matrix elements connecting large and small components, as e.g. $\vec{\sigma}$, β. An even operator commutes with β.

An "odd" operator, however, contains nonvanishing matrix el ements connecting large and small components; it anticommutes with β.

We now discuss the nonrelativistic limit of (1,15), thereby neglecting the small components. In this approximation to order v/c we get the eigenvalue equation

$$H_{NR}\,\varphi = E_{NR}\varphi, \qquad (1,16)$$

with the Hamiltonian of the Pauli-theory

$$H_{NR} = \frac{1}{2m}\vec{\sigma}\cdot(\vec{p}-e\vec{A})\,\vec{\sigma}\cdot(\vec{p}-e\vec{A}) + e\,\Phi =$$

$$= \frac{1}{2m}(\vec{p}-e\vec{A})^2 - \frac{e}{2m}(\vec{\sigma}\cdot\vec{B}) + e\,\Phi\ . \qquad (1,17)$$

I t is of course possible to evaluate higher orders in v/c by this method; no eigenvalue equation, however, results and the Hamiltonian ceases to be Hermitian. Foldy and Wouthuysen therefore suggested another method by which Dirac's theory can be approximated to any order in v/c by means of a two-component theory [4].

2. The DFW Transformation (Free Case)

Foldy and Wouthuysen [5] found a unitary transformation for the diagonalization of the Hamilton-operator. Then the Dirac equation decouples into two-component equations, one for positive and one for negative energy. For free particles large and small components are completely decoupled to any order in v/c; the transformation can be given in closed form.

This unitary transformation has the form[+]

$$\varphi(p) = U(p)\,\psi(p), \tag{1,18}$$

with

$$U(p) = \exp\left\{\frac{\beta}{2} \arc\tan\left(\frac{\vec{\alpha}\cdot\vec{p}}{m}\right)\right\} = \frac{E+m+(\vec{\gamma}\cdot\vec{p})}{\sqrt{2E\,(E+m)}}\,. \tag{1,19}$$

From (1,19) it can be seen immediately that

$$U^{+}(p) = U(-p) = U^{-1}(p)\,. \tag{1,20}$$

The Dirac equation (1,9)

$$P_0\psi = H_0\psi = (\vec{\alpha}\cdot\vec{p} + m\beta)\,\psi,$$

then is transformed into

$$P_0\varphi = H'_0\varphi, \tag{1,21}$$

where

$$H'_0 = U(p)\,H_0 U^{+}(p) = \beta\sqrt{p^2 + m^2}\,. \tag{1,22}$$

Since this transformed Hamiltonian commutes with β,

[+] The last expression is found with an expansion of arc tg $\frac{\vec{\alpha}\vec{p}}{m}$ and of exp $\{/\}$ with the result that $U(p) = \cos\left(\frac{1}{2} \text{ arc tg } \frac{p}{m}\right) + \frac{\vec{\gamma}\vec{p}}{p}\sin\left(\frac{1}{2} \text{ arc tg } \frac{p}{m}\right)$ which gives with tg $\omega = \frac{p}{m}$ and $\cos\omega = \frac{m}{E}$ the final form.

$[H_0', \beta] = 0$, then $\frac{1}{2}(1 \pm \beta)$ are projection operators for states of positive or negative energy respectively, where two components vanish identically in each case. Therefore the resulting equations involve two components only:

$$\varphi = \binom{\varphi_+}{\varphi_-}, \quad p_0 \varphi_\pm = \pm E\, \varphi_\pm . \tag{1,23}$$

The relevant equations decompose further still since also σ_3 commutes with H_0', and thus $\frac{1}{2}(1 \pm \sigma_3)$ again represents a projection operator.

In order to get a better understanding of the transformation obtained it is convenient to discuss the new wave function in configuration space, as has been done by Foldy and Wouthuysen. We can decompose any wave function as

$$\psi(\vec{x}) = \int u(\vec{p}\,')\, e^{i\vec{p}\,'\cdot\vec{x}}\, d^3\vec{p}\,' = \psi_+(\vec{x}) + \psi_-(\vec{x});$$

$$\psi_+(\vec{x}) = \int \tfrac{1}{2}\left(1 + \frac{\beta m + \vec{\alpha}\cdot\vec{p}\,'}{E_{p'}}\right) u(\vec{p}\,')\, e^{i\vec{p}\,'\cdot\vec{x}}\, d^3\vec{p}\,', \tag{1,24}$$

$$\psi_-(\vec{x}) = \int \tfrac{1}{2}\left(1 - \frac{\beta m + \vec{\alpha}\cdot\vec{p}\,'}{E_{p'}}\right) u(\vec{p}\,')\, e^{i\vec{p}\,'\vec{x}}\, d^3\vec{p}\,' .$$

Here ψ_+ represents the wave function for positive, ψ_- the one for negative energy. We now perform the DFW transformation:

$$\varphi_+(\vec{x}) = U\psi_+(\vec{x}) =$$

$$= \left(\frac{1+\beta}{2}\right) \int \sqrt{\frac{E_{p'}}{2(E_{p'}+m)}}\left[1 + \frac{\beta m + \vec{\alpha}\cdot\vec{p}\,'}{E_{p'}}\right] u(\vec{p}\,')\, e^{i\vec{p}\,'\cdot\vec{x}} d^3 p' ;$$

$$\varphi_-(\vec{x}) = U\psi_-(\vec{x}) =$$

$$= \left(\frac{1-\beta}{2}\right) \int \sqrt{\frac{E_{p'}}{2(E_{p'}+m)}}\left[1 - \frac{\beta m + \vec{\alpha}\cdot\vec{p}\,'}{E_{p'}}\right] u(\vec{p}\,')\, e^{i\vec{p}\,'\cdot\vec{x}}\, d^3\vec{p}\,' . \tag{1,25}$$

From this we clearly see that the upper components in (1,25) correspond to positive, the lower ones to negative energy.

Inserting the inverse Fourier transform

$$u(\vec{p}') = \frac{1}{(2\pi)^3} \int \psi(x') \, e^{-i\vec{p}' \cdot \vec{x}'} \, d^3\vec{x}' ,$$

we get the relation between old and transformed wave function

$$\varphi(\vec{x}) = \int K(\vec{x}, \vec{x}') \; \psi(\vec{x}') \, d^3\vec{x}' = \varphi_+(\vec{x}) + \varphi_-(\vec{x}) \tag{1,26}$$

where

$$K(\vec{x}, \vec{x}') = \frac{1}{(2\pi)^3} \int \sqrt{\frac{E_{p'}}{2(E_{p'} + m)}} \, [1 + \\ + \frac{m + (\vec{\gamma} \cdot \vec{p}')}{E_{p'}}] \exp\{i\vec{p}' \cdot (\vec{x} - \vec{x}')\} \, d^3\vec{p}' . \tag{1,27}$$

Because of the momentum dependence of the DFW transformation the kernel $K(\vec{x}, \vec{x}')$ exhibits in its spatial dependence no δ-function; the transformation is not a point-transformation. Had the wave-function in the old representation been localized in a point the transformed wave-function would be smeared out over a finite range (of the Compton wavelength's order of magnitude) [6].

An interesting problem is the interpretation of physical quantities in this new representation. In case of the position-operator we ask for the operator $\vec{X}$ of the old representation which corresponds in the DFW representation to the usual position operators $\vec{x}$, and find with $[x_i, f(p)] = i \frac{\partial}{\partial p_i} f(p)$

$$\vec{X} = U^+ \vec{x} \, U = \vec{x} + \frac{i\beta\vec{\alpha}}{2E} - \frac{i\beta(\vec{\alpha} \cdot \vec{p})\vec{p} + (\vec{\sigma} \times \vec{p})E}{2E^2(E+m)} . \tag{1,28}$$

The time derivative in the new representation is

$$\frac{d}{dt}\vec{x} = i\,[H_0', \vec{x}] = -\,i\,[\vec{x}, \beta E] = \beta\frac{\vec{p}}{E} \, , \tag{1,29}$$

and in the old representation

$$\frac{d\vec{X}}{dt} = i\,[H, \vec{X}] = U^{+}\frac{d}{dt}\vec{x}\,U = \frac{\vec{p}}{E}\,\frac{m\beta + \vec{\alpha}\vec{p}}{E} \, . \tag{1,30}$$

The operator $\frac{1}{E}(m\beta + \vec{\alpha}\vec{p})$ - applied to a positive or negative energy wave function - has the value $+1$ or -1 respectively. So we get the result that the velocity operator for positive energy states is $+\frac{p}{E}$ in either representation.

The operator $\vec{x}$ has been interpreted by Newton and Wigner [7] as the position operator appropriate for describing localized states. It has the properties

$$[X_i, X_j] = 0 \quad ; [X_i, p_j] = i\,\delta_{ij} \tag{1,31}$$

which can be verified in a way similar to (1,30).

3. The Foldy-Heisenberg (F-H) Picture

In addition Foldy and Wouthuysen have developed a method which enables a step-by-step diagonalization of the Hamiltonian in any arbitrary order of v/c or 1/m. This method is important when interactions are present.

The Dirac-Hamiltonian can be decomposed into an even and odd part:

$$H = m\beta + \vec{\alpha}\vec{p} = m\beta + \epsilon + o\ ; \tag{1,32}$$

$$\{\beta, o\} = 0,\ [\beta, \epsilon] = 0\ .$$

These odd and even parts we can write

$$o = \frac{\beta}{2}[\beta, H] = \tfrac{1}{2}(H - \beta H \beta)\ ,$$

$$\epsilon = \frac{\beta}{2}(\{\beta, H\} - 2m) = \tfrac{1}{2}(H + \beta H \beta) - \beta m\ . \tag{1,33}$$

We now perform a unitary transformation with

$$U = e^{iS}\ ,$$

where

$$S = -\frac{i\beta o}{2m} = -\frac{i}{4m}(\beta H - H\beta)\ ; \tag{1,34}$$

so that

$$\psi' = U\psi,\ H' = U\left(H - i\frac{\partial}{\partial t}\right)U^{+}\ .$$

The new Hamiltonian H' then has the form

$$H' = e^{iS}\left(H - i\frac{\partial}{\partial t}\right)e^{-iS} =$$

$$= H + i[S, H] - \tfrac{1}{2}[S, [S, H]] - \frac{\partial S}{\partial t} - \frac{i}{2}\left[S, \frac{\partial S}{\partial t}\right] \ldots \tag{1,35}$$

where we took into account a possible explicit dependence of S on time. Evaluation of the terms leads to

$$H' = \beta m + \epsilon + \mathcal{O} + \frac{1}{2m}[\beta\mathcal{O}, \beta m] + \frac{1}{2m}[\beta\mathcal{O}, \epsilon] +$$
$$+ \frac{1}{2m}[\beta\mathcal{O}, \mathcal{O}] + \frac{1}{8m^2}[\beta\mathcal{O}, [\beta\mathcal{O}, \beta m]] + \ldots \tag{1,36}$$
$$= \beta m + \epsilon + \frac{1}{2m}\beta\mathcal{O}^2 + \frac{1}{2m}\beta[\mathcal{O}, \epsilon] + \frac{i\beta\dot{\mathcal{O}}}{2m} \pm \ldots,$$

and if we decompose H' in analogy to (1,32) we get

$$H' = \beta m + \epsilon' + \mathcal{O}',$$

where

$$\epsilon' = \epsilon + \frac{1}{2m}\beta\mathcal{O}^2, \tag{1,37}$$
$$\mathcal{O}' = \frac{1}{2m}\beta[\mathcal{O}, \epsilon] + \frac{i\beta\dot{\mathcal{O}}}{2m} + \ldots\ldots\ldots .$$

We see that $\mathcal{O}'$ now is at least one order in m^{-1} smaller than $\mathcal{O}$. Further reduction of H' is accomplished by the next DFW transformation

$$S' = -\frac{i\beta}{2m}\mathcal{O}', \tag{1,38}$$

in accordance with the above prescription. By means of sufficiently many transformations and expansion up to the desired order we can diagonalize H apart from terms of higher order in m^{-1}. From the first step, however, we may see already that the computational effort involved in these successive transformations increases enormously for higher orders in m^{-1}.

Let us now come back to the Dirac equation with the interaction term. In closed form no unitary transformation in general can be given which diagonalizes the Hamiltonian. The method of successive transformations, however, can be taken over and H has been diagonalized up to order m^{-4} in case of an interaction with an external field [8]. Here we want to show the procedure up to order m^{-2}.

The Hamiltonian is given in the general form

$$H = H_o + H_{int} = \vec{\alpha}\cdot\vec{p} + \beta m + \beta J_\mu A^\mu, \tag{1,39}$$

where $A_\mu = (\Phi, \vec{A})$ represents the unquantized electromagnetic field. For the electromagnetic current (e.g. of the electrons) we take the ansatz:

$$J_\mu = e\,\gamma_\mu\,, \tag{1,40}$$

i.e. we are discussing a point-particle without anomalous magnetic moment. (The general formula with anomalous moment and form factors one can find in [9].)

With this the Hamiltonian (1, 39) becomes

$$\begin{aligned} H &= \vec{\alpha}\cdot(\vec{p}-e\vec{A}) + \beta m + e\Phi \\ &= \beta m + \epsilon + \mathfrak{o}\,; \end{aligned} \tag{1,41}$$

where

$$\mathfrak{o} = \vec{\alpha}\cdot(\vec{p}-e\vec{A})\,,$$

$$\epsilon = e\Phi\,.$$

Performing the successive transformations up to order m^{-2} we find

$$\begin{aligned} H' &= \beta m + \epsilon' = \\ &= \beta\left(m+\frac{\mathfrak{o}^2}{2m}-\frac{\mathfrak{o}^4}{8m^3}\right)+\epsilon-\frac{1}{8m^2}[\mathfrak{o},[\mathfrak{o},\epsilon]]-\frac{i}{8m^2}[\mathfrak{o},\dot{\mathfrak{o}}] = \\ &= \beta\left[m+\frac{\vec{p}^2}{2m}-\frac{e}{2m}(\vec{p}\vec{A}+\vec{A}\vec{p})+\frac{e^2}{2m}\vec{A}^2-\frac{\vec{p}^4}{8m^3}\right]+ \\ &\quad + e\Phi-\frac{e}{8m^2}\operatorname{div}\vec{E}-\frac{e}{2m}\beta\,\sigma\cdot B+\frac{e}{8m^2}[\vec{\sigma}\cdot(\vec{p}\times\vec{E})-\vec{\sigma}\cdot(\vec{E}\times\vec{p})]\,. \end{aligned} \tag{1,42}$$

The physical meaning of the terms shall now be analyzed. The terms in the first bracket have their origin in the expansion of

$$\sqrt{(\vec{p}-e\vec{A})^2+m^2}$$

up to the desired order in m^{-2}; the convection current is $(\vec{p}\vec{A} + \vec{A}\vec{p})$.

The electrostatic interaction is modified by the Darwin term

$$\left(-\frac{e}{8m^2}\,\mathrm{div}\,\vec{E}\right), \tag{1,43}$$

which is connected with the "Zitterbewegung". Since the particle (electron) performs oscillations within $\delta r \cong \frac{1}{m}$ it experiences a smeared-out Coulomb potential. The correction to the potential energy can be easily estimated, it agrees with the Darwin term in the order of magnitude:

$$<\delta V> = <V(\vec{r} + \delta\vec{r})> - <V(\vec{r})> =$$

$$= <\delta r \frac{\partial V}{\partial r} + \tfrac{1}{2} \sum_{i,j} \delta r_i \, \delta r_j \frac{\partial^2 V}{\delta r_i \, \delta r_j}> \cong$$

$$\cong \frac{1}{6}\,\delta r^2 \cdot (\vec{\nabla}^2 V) \cong -\frac{1}{6m^2}\,\mathrm{div}\,\vec{E}.$$

The term $\left(-\frac{e}{2m}\,\beta\vec{\sigma}\cdot\vec{B}\right)$ describes the coupling between spin and magnetic field $\vec{B}$, as already known from the Pauli equation. Finally, the last term represents the spin-orbit coupling. In a spherical symmetric static potential (rot $\vec{E} = 0$) it takes the well known form

$$H_{\text{spin-orbit}} = \frac{e}{8m^2}\,\frac{1}{r}\,\frac{\partial V}{\partial r}\,(\vec{\sigma}\cdot\vec{L}), \tag{1,44}$$

since

$$\sigma \cdot (\vec{E} \times \vec{p}) = -\frac{1}{r}\,\frac{\partial V}{\partial r}\,\vec{\sigma}\,(\vec{r} \times \vec{p}) = -\frac{1}{r}\,\frac{\partial V}{\partial r}\,\vec{\sigma}\cdot\vec{L}\,.$$

The factor of $(1/8\,m^2)$ in $H_{\text{spin-orbit}}$ also takes into account the contribution due to the Thomas precession. Neuer and Urban, employing some different arguments, also derived the terms in H' and discussed their physical meaning in detail [10].

Since in this procedure the total Hamiltonian $H = H_o + H_{int}$ has been diagonalized, at least to some approximation, we shall call this the "Foldy-Heisenberg" picture. It is applicable if the interaction with an external field (not quantized), e.g. with an external magnetic field or the Coulomb field of a nucleus, is considered only. In the latter case spin-orbit and Darwin term lead to the correct energy levels of the hydrogen atom.

4. The Foldy-Interaction (F-I) Picture

We experience a totally different situation in case of an interaction with a quantized field, e.g. the Møller-potential of a scattering electron, although the formalism of the DFW transformation has been applied to such cases, too. The transformed Hamiltonian (1,42) then contains terms proportional to A_μ^2 or even higher powers; exactly those cause violation of energy-momentum conservation and therefore have to be omitted [11] since a single photon, which has been emitted by an electron, cannot possibly be absorbed twice, e.g. by the scattered electron. Thus the situation seems quite unsatisfactory.

Breitenlohner [12] then proposed another procedure which leads to the same results but allows a consistent interpretation of the scattering process: Not the total Hamiltonian H, as previously, but only the free Hamiltonian H_o becomes diagonalized. Therefore this representation has been called "Foldy-Interaction" picture.

The starting point again is the Schrödinger equation for a Dirac particle (see sections 2 and 3):

$$i\dot{\psi} = H\psi = (H_o + H_{int})\psi = (\vec{\alpha}\cdot\vec{p} + \beta m + \beta J_\mu A^\mu)\psi . \qquad (1,45)$$

By means of the unitary transformation

$$U(p) = \frac{E + m + (\vec{\gamma}\cdot\vec{p})}{\sqrt{2E(E+m)}} \;, \tag{1,46}$$

the free Hamiltonian H_o can be diagonalized to

$$H'_o = U\, H_o\, U^+ = \beta E \;,$$

$$E = +\sqrt{\vec{p}^{\,2} + m^2}\,. \tag{1,47}$$

With this fixed transformation into the Foldy-Interaction picture we get for the new Hamiltonian

$$H' = UHU^+ = H'_o + H'_{int} = \beta E + UH_{int}\, U^+ \;, \tag{1,48}$$

$$\varphi = U\,\psi\;.$$

Now, of course, the transformed interaction term is linear in A_μ. In addition, the transformation (1,48) can be done in closed form, therefore it is not necessary to employ a series expansion in powers of m^{-1} whose convergence properties are not easily intelligible. However, H' contains now "odd" terms.

If we consider a scattering process we are interested in the matrix elements of the interaction Hamiltonian between free states with initial momentum $\vec{p}_1$ and final momentum $\vec{p}_2$ respectively:

$$\psi^+(\vec{p}_2)\, H_{int}\, \psi(\vec{p}_1) = \varphi^+(\vec{p}_2)\, H'_{int}\, \varphi(\vec{p}_1) =$$

$$= \varphi^+(\vec{p}_2)\, U(\vec{p}_2)\, H_{int}\, U^+(\vec{p}_1)\, \varphi(\vec{p}_1)\;. \tag{1,49}$$

The expressions for

$$U(p_2)\, 1\, U^+(p_1),\; U(p_2)\, \vec{\alpha}\, U^+(p_1) \text{ etc.}$$

we shall not list explicitly (see[12]), the result being that the transformed Hamiltonian H'_{int} consists of four 2×2 matrices corresponding to the processes: particle scattering, antiparticle scattering, pair production, pair annihilation (the last two parts are not contained in the Foldy-Heisenberg formalism):

$$H'_{int} = \left[\begin{array}{l|l} +E_1 \to +E_2 & -E_1 \to +E_2 \\ \text{particle} & \text{pair} \\ \quad\text{scattering} & \quad\text{production} \\ \hline +E_1 \to -E_2 & -E_1 \to -E_2 \\ \text{pair} & \\ \text{annihilation} & \text{antiparticle} \\ & \quad\text{scattering} \end{array}\right] = \left[\begin{array}{c|c} \text{even} & \text{odd} \\ \hline \text{odd} & \text{even} \end{array}\right] \qquad (1,50)$$

Finally we want to evaluate H'_{int} in case that the electro-magnetic current is given by

$$J_\mu = eF_1\,\gamma_\mu + \frac{e}{4m} F_2\,[\not{q}, \gamma_\mu] , \qquad (1,51)$$

where F_1 and F_2 are Dirac and Pauli form factors respectively. With the abbreviations

$$q = p_2 - p_1, \quad P = p_1 + p_2 ;$$

$$N_i = [\,2E(\vec{p}_i)\,(E(\vec{p}_i) + m)]^{-\frac{1}{2}}, \quad A = (E(\vec{p}_2) + m)\,(E(\vec{p}_1) + m) , \qquad (1,52)$$

$$B = \frac{\vec{P}^{\,2} - \vec{q}^{\,2}}{4} , \quad C = \frac{E(\vec{p}_1) + E(\vec{p}_2)}{2} + m, \quad D = \frac{E(\vec{p}_2) - E(\vec{p}_1)}{2}$$

the transformed interaction Hamiltonian has the form

$$H'_{int} = eN_1 N_2 \,\{[\,(2mC + \frac{\vec{P}^2}{2})\,F_1 + (2mC - \frac{\vec{q}^{\,2}}{2m}\,C + \frac{(\vec{P}\cdot\vec{q})}{2m}.\,D - A +$$

$$+ \frac{\vec{P}^2 + \vec{q}^{\,2}}{4}\,)\,F_2\,]\,\varphi + [\,-CF_1 + (\frac{1}{2m}\,(A-B) - C\,)\,F_2\,]\,\beta\,(\vec{P}\cdot\vec{A}\,) -$$

$$- i\,[\,\tfrac{1}{2}\,F_1 + \tfrac{1}{2}\,CF_2\,]\,(\vec{\sigma}, \vec{P}, \vec{q}\,)\,\Phi + \frac{i}{4m^2}\,F_2\,\beta(\vec{\sigma}, \vec{P}, \vec{q}\,)\,(\vec{P}\cdot\vec{A}\,) +$$

$$+ i\beta (F_1 + F_2) [D (\vec{\sigma}, \vec{P}, \vec{A}) - C (\vec{\sigma}, \vec{q}, \vec{A})] \} +$$

$$+ eN_1 N_2 \{ \tfrac{i}{2} (F_1 + F_2) (\vec{P}, \vec{q}, \vec{A}\ ) \gamma_5 + [mF_1 + (\frac{1}{2m} A - C -$$

$$- \frac{\vec{P}^2 + \vec{q}^2}{8m}) F_2] (\vec{\gamma} \cdot \vec{q}) \Phi + [- DF_1 + \frac{(\vec{P} \cdot \vec{q})}{4m} F_2] (\vec{\gamma} \cdot \vec{P}) \Phi - \qquad (1,53)$$

$$- (A + B) (F_1 + F_2) (\vec{\alpha} \cdot \vec{A}) + \frac{1}{2m} DF_2 (\vec{\alpha} \cdot \vec{q}) (\vec{P} \cdot \vec{A}) +$$

$$+ [\tfrac{1}{2} F_1 + \tfrac{1}{2} F_2 - \frac{1}{2m} C F_2] (\vec{\alpha} \cdot \vec{P}) (\vec{P} \cdot \vec{A}) \} .$$

Expansion of H'_{int} in powers of m^{-1} up to second order results in

$$H'_{int} = e \{ (1 + \frac{q^2}{8m^2}) F_E \Phi - F_E \frac{1}{2m} \beta (\vec{P} \cdot \vec{A}) - \frac{i\beta}{2m} F_M (\vec{\sigma}, \vec{q}, \vec{A}) +$$

$$+ \frac{i}{8m^2} (F_E - 2F_M) (\vec{\sigma}, \vec{P}, \vec{q}) \Phi \} + \text{odd terms} ; \qquad (1, 54)$$

where

$$(1 + t) F_E = G_E = F_1 - F_2 t ,$$

$$(1 + t) F_M = G_M = F_1 + F_2 , \qquad (1,55)$$

$$t = - \frac{q_\mu q^\mu}{4m^2} .$$

Here G_E and G_M are the electric and magnetic form factors as defined by [13].

We finally want to remark that the expression (1,54) for H'_{int} exactly reproduces the terms of the Foldy-Heisenberg Hamiltonian to order m^{-2}, which are linear in the electromagnetic interaction. This connection therefore validates in scattering problems the long used method of just considering the Hamiltonian in the Foldy-

Heisenberg picture. In (1,54) the second term is again the convection current, the third the interaction of the magnetic moment and the latter is the spin-orbit term. The first term is the electrostatic interaction corrected by the Darwin-term.

5. Connection between DFW and Lorentz Transformation

The connection between the DFW transformation and the Lorentz-transformation from laboratory to rest system has been shown by Good [14]. We consider a Lorentz transformation to a system x_L, moving with velocity $\vec{v} = \vec{p}/E$ relative to the original system x_R:

$$x_L^\mu = a^\mu_\nu x_R^\nu \, , \quad a^\nu_\mu a^\mu_\rho = \delta^\nu_\rho, \quad \det|a| = 1 \, ; \qquad (1,56)$$

with

$$a_{oo} = \frac{E}{m}, \quad a_{ij} = \delta_{ij} + \frac{P_i P_j}{m(E+m)} \, , \quad a_{ok} = -\frac{P_k}{m} .$$

The wave function ψ (x) then transforms according to

$$\psi_L (x) = \Lambda \, \psi_R (x_R) \, ,$$

$$\Lambda = \frac{E + m + \vec{\alpha}\cdot\vec{p}}{\sqrt{2m(E+m)}} \, . \qquad (1,57)$$

Here Λ is the so-called "boost" operator.

We now define the generalized operator

$$\Lambda(\vec{p}) = \frac{|P_o| + m + \vec{\alpha}\cdot\vec{p}\,(P_o/\,|P_o|)}{\sqrt{2m\,(\,|P_o| + m)}} \, , \quad P_o = i\frac{\partial}{\partial t} \, ; \qquad (1,58)$$

applied to a plane wave state it reproduces the operator of the Lorentz transformation in (1,57). In (1,58) P_o can be substituted by the Hamiltonian H_o if it is applied directly to a wave function (P_o commutes with Dirac matrices whereas H_o does not).

If we now consider a state $\Lambda^{-1}(\vec{p})\,\psi_L$, where ψ_L stands for an arbitrary Dirac wave function, we see that

$$\Lambda^{-1}(\vec{p})\,\psi_L = \frac{|P_o| + m - \vec{\alpha}\cdot\vec{p}\,(P_o/|P_o|)}{\sqrt{2m\,(|P_o| + m)}}\,\psi_L$$

$$= \frac{|H_o| + m - [(\vec{\alpha}\cdot\vec{p})/|H_o|\,]\,H_o}{\sqrt{2m\,(|H_o| + m)}}\,\psi_L\ .$$

Since

$$H_o\,\psi_L = (\vec{\alpha}\cdot\vec{p} + \beta m)\,\psi_L\,,$$

we get

$$\Lambda^{-1}(\vec{p})\,\psi_L = \sqrt{m/\,|H_o|}\ \frac{|H_o| + m + \beta\,\vec{\alpha}\cdot\vec{p}}{\sqrt{2\,|H_o|\,(|H_o| + m)}}\ \psi_L =$$

$$= \sqrt{m/E}\ \frac{E + m + \vec{\gamma}\cdot\vec{p}}{\sqrt{2E\,(E+m)}}\ \psi_L = \sqrt{m/E}\ U(\vec{p})\,\psi_L =$$

$$= \sqrt{m/E}\ \psi_{DFW}\ ;$$

or

$$\psi_{DFW} = \sqrt{E/m}\ \psi_R\ . \tag{1,59}$$

The DFW transformed wave function thus differs from the wave function in the rest system only by the normalization factor $\sqrt{E/m}$. The origin of this factor is due to the fact that ψ_{DFW} is normalized in the laboratory system and ψ_R in the rest system:

$$\psi_L^*\,\psi_L = \frac{E}{m}\ \psi_R^*\,\psi_R\ . \tag{1,60}$$

In order to further clarify the connection we write the DFW operator $U(\vec{p})$ in the form:

$$U(p) = \sqrt{\frac{E}{m}} \left\{ \frac{E+m-\vec{\alpha}\cdot\vec{p}}{\sqrt{2m(E+m)}} \cdot \frac{E+\vec{\alpha}\cdot\vec{p}+\beta m}{2E} + \right.$$

$$\left. + \frac{E+m+\vec{\alpha}\cdot\vec{p}}{\sqrt{2m(E+m)}} \cdot \frac{E-\vec{\alpha}\cdot\vec{p}-\beta m}{2E} \right\} = \qquad (1,61)$$

$$= \sqrt{E/m}\, \{ \Lambda(-\vec{p}) \frac{E+H_o}{2E} + \Lambda(\vec{p}) \frac{E-H_o}{2E} \} .$$

We now clearly see that, apart from the relativistic normalization factor $\sqrt{E/m}$, the operator U consists of products of projection operators for states with $p_o = H_o = \pm E$ and Lorentz transformations, which transform the vectors $(\pm E, \vec{p})$ to the relevant rest system $(\pm m, 0)$. It is thus intelligible that the DFW transformation is equivalent to a transformation to the rest system. This connection, however, is lost in the Foldy-Heisenberg picture which leads to ambiguities as discussed in section 7.

6. Connection between F-H and F-I Picture

If we apply the iteration procedure (section 3) to the total Hamiltonian (1,45) of the F-I representation, with βE instead of βm, we get the F-H Hamiltonian as a power series in $\vec{p}/E$ instead of $\vec{p}/m$. A closer inspection shows that the resulting terms, linear in the interaction, exactly coincide with the even part of the F-I Hamiltonian. The agreement which we found at the end of section 4 therefore was not accidental but holds to any order in m^{-1}. Now, however, correct relativistic kinematics are introduced since the F-I transformation is not a semirelativistic approximation but rather a transformation to the rest system; secondly, an expansion in $\vec{p}/E$ certainly converges better than one in $\vec{p}/m$, especially for greater momenta and momentum transfer. Furthermore the connection between these two pictures puts the iteration procedure of the F-H picture onto more solid grounds.

7. Ambiguities of the F-H Picture

Since the unitary transformation of the F-H picture is fixed only by the requirement that boxes constituting the total Hamiltonian should be diagonalized (i.e. it should commute with β) its form is not unambiguous. It is always possible to apply arbitrary unitary transformations to the two-dimensional subspaces. Eriksen [15] investigated several versions of closed form transformations and indeed found different results. The difference is due to momentum dependent rotations which are different for particles and antiparticles. Since the spin operators have to be transformed accordingly the matrix elements between the same physical states remain of course unchanged.

Let us consider the DFW transformed interaction term of a point particle without anomalous magnetic moment in an electromagnetic field (up to order m^{-2}) :

$$H'_{int} = e\,(1-\frac{\vec{q}^{\,2}}{8m^2})\,\Phi - e\,\frac{\beta}{2m}(\vec{p}\vec{A} + \vec{A}\vec{p}) - e\,\frac{\beta}{2m}(\vec{\sigma}\cdot\vec{B}) + \frac{e}{8m^2}\,(\vec{\sigma}\cdot(\vec{p}\times\vec{E} - \vec{E}\times\vec{p}))\ . \tag{1,62}$$

We transform now (1,62) by means of

$$V(\vec{p}) = \exp\,\{\frac{i\epsilon\beta}{2m}\,(\vec{\sigma}\cdot\vec{p})\}\ , \tag{1,63}$$

to get the transformed Hamiltonian as

$$H''_{int} = V\,H'_{int}\,V^{+} = e\,[\,1-(1+\epsilon^2)\,\frac{q^2}{8m^2}\,]\,\Phi - e\,\frac{\beta}{2m}\,(\vec{p}\vec{A} + \vec{A}\vec{p}) - e\,\frac{\beta}{2m}\,(\vec{\sigma}\cdot\vec{B}) - e\,\frac{\epsilon\beta}{2m}\,(\vec{\sigma}\cdot\vec{E}) + (1+\epsilon^2)\,\frac{e}{8m^2}\,(\vec{\sigma}\cdot(\vec{p}\times\vec{E}-\vec{E}\times\vec{p}))\ . \tag{1,64}$$

We see that now both the Darwin and the spin-orbit terms are multiplied by other numerical factors, and an electric moment of magnitude $\epsilon/2m$ seems to be added. Of course it is no real electric moment and causes no violation of parity, since $\vec{\sigma}$ no longer is the spin operator. The choice $\epsilon = -1$ reproduces one of the results of Eriksen and Kolsrud [15].

8. Application of the DFW Transformation

The most important applications of this method are the hydrogen atom and electron scattering processes. The energy levels of the hydrogen atom have been calculated by means of the Hamiltonian in the F-H picture taken up to order m^{-2} [2].
McVoy and Van Hove [16] used that part of the F-H Hamiltonian, as mentioned earlier, which is linear in the electromagnetic interaction, to calculate cross sections for elastic and inelastic electron scattering processes where the scattered electron was represented by its Møller potential. In an exact treatment of such problems one should use the box in the left hand upper corner of the F-I Hamiltonian; we have shown, however, that these two (F-H and left hand upper box of F-I) coincide. These authors also derived the Rosenbluth formula for elastic scattering from free nucleons. In close connection with this several papers at our Institute have dealt with elastic [11] and inelastic electron-deuteron scattering[+] which is of great importance for the determinations of neutron form factors.

Finally we want to discuss the application of the DFW-transformation to pair-creation and - annihilation processes. Here the odd parts of the F-I Hamiltonian (1,50) are involved. Creation and annihilation of free particles are much easier calculated by using the original Dirac equation and therefore no need for a DFW transformation exists. Also processes like one-quantum-pair annihila-

+ See chapter III.

tion of antiprotons or positrons with bound protons or electrons, where possibly two-component wave functions have to be used, are much too unrealistic (vanishingly small cross sections). The application, however, is not restricted to electromagnetic interactions; for strong or weak interactions one could (assuming sufficient knowledge of the wave functions) calculate processes like

$$\bar{p} + d \rightarrow n + \pi^{0} \; ,$$

$$\bar{p} + d \rightarrow p + \pi^{-} \; ;$$

or relativistic corrections to β-decays like

$$H^{3} \rightarrow He^{3} + e^{-} + \bar{\nu} \; , \text{ etc.}$$

Literature

[1] P.A.M. Dirac, Proc. Cambridge Phil. Soc. 30, 150 (1934).

[2] J.D. Bjorken, S.D. Drell, Relativistic Quantum Mechanics, New York: McGraw-Hill (1964).

[3] J.D. Bjorken, S.D. Drell, Relativistic Quantum Fields, New York: McGraw-Hill (1965).

[4] A. Messiah, Quantum Mechanics, Vol. II (1964).

[5] L.L. Foldy, S.A. Wouthuysen, Phys. Rev. 78, 29 (1950).

[6] M.E. Rose, Relativistic Electron Theory, New York: John Wiley & Sons, Inc. (1961)

[7] T.D. Newton, E.P. Wigner, Rev. Mod. Phys. 21, 400 (1949).

[8] M. Baktavatsalov, J. Phys. Radium (Paris) 20, 159 (1961).

[9] K. Hoelzl, G. Saller, P. Urban, Acta Phys. Austr. 19, 168 (1964).

[10] H. Neuer, P. Urban, Acta Phys. Austr. 15, 380 (1962).

[11] H. Neuer, P. Urban, Acta Phys. Austr. 16, 60 (1963).

[12] P. Breitenlohner, Acta Phys. Austr. 22, 217 (1966).

[13] F.J. Ernst, R.G. Sachs, K.C. Wali, Phys. Rev. 119, 1105 (1960).

[14] R.H. Good, Jr., M.E. Rose, Nuovo Cim. 24, 864 (1962).

[15] E. Eriksen, M. Kolsrud, Nuovo Cim., Suppl. 18, 1 (1960).

[16] K.M. McVoy, L. van Hove, Phys. Rev. 125, 1034 (1962).

II. Determination of Proton-Form Factors Derived from Electron-Proton-Scattering

1. Introductory Remarks

Following a series of outstanding successes of QED, in other words the interaction of the electromagnetic field with the field of electrons and positrons, achieved during the last years of the forties, scientists began to describe the electromagnetic qualities of strong-interacting particles in connection with this new formalism. Especially striking and demonstrative was the calculation of the scattering of electrons on protons. This process, in analogy to the Møller-scattering, was attributed to the interaction of the charge and the magnetic moments, both of which are assumed as a point, with the external electromagnetic field $A_\mu^{(e)}(x)$ (the so-called Møller-potential) of the scattered electrons. In the course of the experiments it was found that calculated cross sections for momentum transfers of several Fermi^{-1} ($1f^{-1} = 200$ MeV) were decidedly greater than previous experimental values. The cause of such a discrepancy is to be found in the fact that in the matrix element

$$\int d^4x\, J^\mu(x)\, A_\mu^{(e)}(x)$$

we have substituted the QED expression i.e. the expression for a point particle for the proton-current $J^\mu(x)$. Thus it was essential to consider both the anomalous magnetic moment of the proton and an apparent spatial extension of charge and moment. It was possible to give a clear and instructive interpretation of these corrections in the following explanations: It has been established that resulting from the strong coupling of nucleons with other strong interacting particles the nucleon is surrounded by a cloud of such virtual particles which are usually identified with known mesons. Consequently further contributions of meson-currents or contributions of virtual nucleon-antinucleon pairs have to be added for a comprehensive description of the electron-proton-scattering. These contributions would have to take into account also a possible interaction of the electrons with one of these trabants instead of a direct interaction with the bare proton. An extensive calculation of these contributions has been the subject matter of a large number of field-theoretical papers. Particularly some dispersion-theoretical reports have given essential indications as to predominant interaction-mechanisms and, finally, even led to a successful prediction of the vector-mesons, a so far unknown group of elementary particles. Thus the electron-scattering experiments enabled us to check different models of nucleon structure. It was therefore necessary to transform the respective scattering formulas into another expression which should enable us to interpret the experimental cross-sections of the spatial charge and moment distribution of the nucleons in a direct way. This aim was effected by applying a more general ansatz for a nucleon current: The point-interaction was replaced by a spatially smeared out interaction and, moreover, the so-called Pauli-term for the anomalous magnetic moment of the proton was included.

We shall see that the improved current-operator yields scalar form factors the Fourier transforms of which represent the spatial charge and moment distribution of the proton in a special Lorentz-frame in which the time-component of the four-momentum-transfer q_μ vanishes. The problem is not really the distribution in the actually interesting frame, in other words the rest frame of the nucleons, but we are inclined to assume that the true structure of the particle at rest does not differ from the structure deduced from the form factors in any particular way. A primary condition for this theory is: in the static case $q^2 \to 0$ the form factors are to take such values that the effective charge in the current-operator will take the value of the renormalized observed charge of the nucleon; the magnetic part is reduced to the total (or anomalous) magnetic moment of the nucleons. The following section describes the derivation of the Rosenbluth-formula for e-p-scattering, in section 3 the proton-form factors (determined by the electron-scattering-experiments) are indicated and discussed.

2. Relativistic Electron-Proton-Scattering (Rosenbluth Formula)

We now calculate relativistic e-p-scattering in the lab-system with k and k' = k-q describing the four-momentum of the ingoing and outgoing electron respectively. The electron mass may be neglected with respect to the electron-energy at the energies considered herein (100 MeV and more) with regard to all kinematic relations. The proton-momenta are denoted by p and p' before and after the scattering.

Thus the corresponding Feynman diagram has the following form (higher powers in e^2 are neglected for the present energies (Karzas [1])).

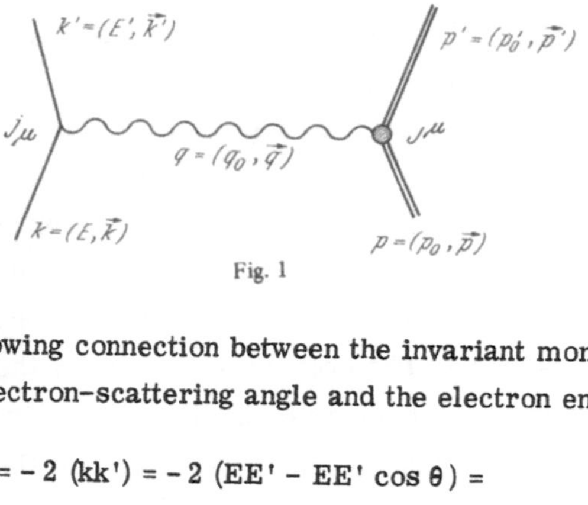

Fig. 1

We find the following connection between the invariant momentum transfer, the electron-scattering angle and the electron energies:

$$q^2 = (k-k')^2 = -2\,(kk') = -2\,(EE' - EE'\cos\theta) =$$

$$= -4\,EE'\sin^2\frac{\theta}{2}\,. \qquad (2,1)$$

From

$$q^2 = q_0^2 - \vec{q}^{\,2} = (E-E')^2 + 4\,EE'\sin^2\frac{\theta}{2} \qquad (2,2)$$

and the energy-momentum-conservation we derive

$$E' = E\,(1 + \frac{2E}{m}\sin^2\frac{\theta}{2})^{-1} \qquad (2,3)$$

and

$$q^2 = -4E^2 \sin^2\frac{\theta}{2}\,(1 + \frac{2E}{m}\sin^2\frac{\theta}{2})^{-1} \;. \qquad (2,1a)$$

Instead of a point interaction (we use in the calculation the Heisenberg picture)

$$\int \frac{d^4q}{(2\pi)^4} <p'\;\Big|\int J^{\mu}(x)\,e^{-iqx}\,d^4x\,\Big|\,p> A_{\mu}^{(e)}(q) \qquad (2,4)$$

with

$$A_{\mu}^{(e)}(q) = \frac{1}{q^2} <k'\;\Big|\int j_{\mu}^{e}(x)\,e^{iqx}\,d^4x\,\Big|\,k> \,,$$

we assume in the transition element a spatially smeared out coupling of the virtual photon with the electromagnetic p-e-system. Thus

$$M_{fi} = \int d^4x' \int d^4x <p'\;\Big|J^{\mu}(x)\,e^{-iqx'}\,F(x'-x)\,\Big|\,p> A_{\mu}^{(e)}(x') \;.$$

This ansatz is to correspond to an inner structure of the p-γ-vertex part, which may be best illustrated in the language of the Feynman diagrams

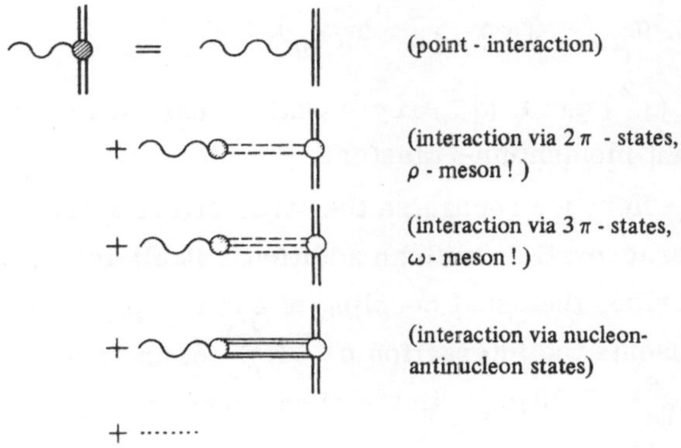

Fig. 2

While we may write the matrix element of the electron current operator following the well-known rules of QED ($\bar{u}u = 2m$ in our normalization)

$$\frac{1}{q^2} \langle k' | \int j_\mu (x) e^{iqx} d^4x | k \rangle =$$

$$= \sqrt{1/4 E E'} \, a_\mu (2\pi)^4 \delta^4 (k' - k + q) \qquad (2,5)$$

with

$$a_\mu = \frac{e}{q^2} (\bar{u}_{k'} [\gamma_\mu] u_k) ,$$

the calculation of the proton-matrix element requires several additional considerations (Drell [2]).

Extracting the exponential factors from the proton-field operators we are led to the matrix element $\langle p' | J^\mu (o) | p \rangle$ of the operator $J^\mu (x)$ for x = o. The use of the Lorentz- and gauge invariance and the validity of the free Dirac equation for the proton spinors in the in- and out-states gives

$$\langle p' | J_\mu (o) | p \rangle = e \sqrt{1/4 p_o p_o'} \, (\bar{u}_{p'} [A (q^2) \gamma_\mu +$$

$$+ \frac{1}{2m} B (q^2) i \sigma_{\mu\nu} q^\nu] u_p) \qquad (2,6)$$

with

$$\sigma_{\mu\nu} = \frac{i}{2} (\gamma_\mu \gamma_\nu - \gamma_\nu \gamma_\mu) ,$$

$A (q^2)$ and $B (q^2)$ representing real, scalar functions of the invariant momentum-transfer.

Here we recognize the structure of a current - operator for a Dirac-particle with an additional Pauli-term. The first term describes the usual coupling of a Dirac-particle, the second one represents the interaction $\sigma^{\mu\nu} F^{(e)}_{\mu\nu}$ of an anomalous magnetic moment $g \frac{e}{2m} = 1{,}79 \, \mu_B$ with the electromagnetic field

$$F^{(e)}_{\mu\nu} = \partial_\nu A^{(e)}_\mu - \partial_\mu A^{(e)}_\nu \ .$$

We shall have to assume different spatial distributions for both mechanisms of interaction. For each of the two terms a space-time integration results as follows (Drell [2])

$$\int d^4x' \int d^4x \, e^{i(p'-p)x} F_{1,2}(x'-x) \, e^{-iqx'}$$

which may be transformed into

$$\int d^4x \, e^{i(p'-p-q)x} F_{1,2}(q^2) \tag{2,7}$$

with

$$F(q^2) = \int d^4x \, e^{-iqx} F(x) \tag{2,8}$$

or

$$F(x) = \frac{1}{(2\pi)^4} \int d^4q \, e^{iqx} F(q^2) \ . \tag{2,9}$$

We identify A and B in the current operator with $F_1(q^2)$ and $F_2(q^2)$ respectively. In a Lorentz frame with a vanishing energy-component of the momentum-transfer $F(q^2)$ equals $F(\vec{q}^{\,2})$. By means of a three-dimensional Fourier-transformation the above results in a spatial distribution of the electric charge (and at the same time of the normal magnetic moment) or in the anomalous magnetic moment in this special frame. Thus there is no direct connection between the form factors gained from electron-scattering-experiments on the one hand and the charge- and moment-distribution of the proton at rest on the other. We must be careful when discussing the electromagnetic structure of the proton or neutron derived from form factor measurements.

Integrating over d^4q we obtain for the transition amplitude the following expression

$$M_{fi} = (\bar{u}_{p'} \, [\gamma_\mu F_1(q^2) +$$

$$+ i\sigma_{\mu\nu} q^{\nu} \frac{F_2(q^2)}{2m}] \mu_p) \, e \, a^{\mu} (16EE'p_o p'_o)^{-\frac{1}{2}} (2\pi)^4 \delta^4 (p'-p+k'-k) \ , \qquad (2,10)$$

in which q_μ results from the four-momentum conservation. Before turning toward the calculation of the cross-section we should like to state a second form for $<p'|J_\mu(o)|p>$.

This form is more satisfactory than the first one for two reasons: first it permits a better physical interpretation, secondly it is better adapted to the latter evaluation of the experimental results. The motivation for the second argument will become clear by considering the form of the transition cross-section. Thus it will be indicated in connection with the evaluation of the form factors from the measured cross section. Concerning the first point, namely the physical interpretation of the individual terms in $J_\mu(o)$, the suggestion was put forward to replace the Dirac and Pauli form factors by the so-called Sachs form factors G_e, G_M (Sachs [3], Ernst [4]). G_E describes the purely electric and G_M the purely magnetic properties of the nucleons. G_E and G_M are defined as follows:

$$G_E(q^2) = F_1(q^2) - \tau F_2(q^2) \qquad (2,11a)$$

$$G_M(q^2) = F_1(q^2) + F_2(q^2) \qquad (2,11b)$$

$$F_1(q^2) = \frac{1}{1+\tau} (G_E(q^2) + \tau G_M(q^2)) \qquad (2,12a)$$

$$F_2(q^2) = \frac{1}{1+\tau} (- G_E(q^2) + G_M(q^2)) \qquad (2,12b)$$

with

$$\tau = - \frac{q^2}{4m^2} \ . \qquad (2,13)$$

Inserting (2,12a) and (2,12b) into

$$e(\bar{u}_{p'}[\gamma_\mu F_1 + i\sigma_{\mu\nu}q^\nu \frac{F_2}{2m}]u_p) \equiv (\bar{u}_{p'}[J_\mu]u_p)$$

we may also split the electromagnetic current of the proton into a purely electric and a purely magnetic portion:

$$J_\mu = e F_1 X_\mu^{(1)} + e F_2 X_\mu^{(2)} = e F_E X_\mu^{(E)} + e F_M X_\mu^{(M)} \quad (2,14)$$

with

$$F_{E,M} = (1+\tau)^{-1} G_{E,M} , \quad (2,15)$$

$$X_\mu^{(1)} = \gamma_\mu , \quad (2,16)$$

$$X_\mu^{(2)} = \frac{1}{4m} (\not{q}\gamma_\mu - \gamma_\mu \not{q}) , \quad (2,17)$$

$$X_\mu^{(E)} = \frac{1}{2m} P_\mu , \quad (2,18)$$

$$X_\mu^{(M)} = \frac{1}{8m^2} (\gamma_\mu \not{P}\not{q} - \not{q}\not{P}\gamma_\mu) , \quad (2,19)$$

$$P = p + p' \quad (2,20)$$

$$q = p' - p . \quad (2,21)$$

The results (2,14) – (2,19) can be easily verified with the help of the Dirac equation in the following way:

$$4m^2 (\bar{u}_{p'} [X_\mu^{(M)}] u_p) =$$

$$= 2(\bar{u}_{p'} [\gamma_\mu ((pp') + m^2) - m P_\mu] u_p) =$$

$$= (\bar{u}_{p'} [\gamma_\mu (4m^2 - q^2) - 2m P_\mu] u_p) .$$

Therefore

$$(\bar{u}_{p'} \gamma_\mu u_p) = \frac{1}{4m^2} \frac{1}{1+\tau} (\bar{u}_{p'} [4m^2 X_\mu^{(M)} + 2m P_\mu] u_p) ;$$

similarly we get

$$(\bar{u}_{p'}[i\sigma_{\mu\nu}q^{\nu}]u_p) = (\bar{u}_{p'}[2m\gamma_\mu - P_\mu]u_p) =$$

$$= \frac{1}{2m}\frac{1}{1+\tau}(\bar{u}_{p'}[4m^2 X_\mu^{(M)} - 2m\tau P_\mu]u_p)$$

and obtain

$$(\bar{u}_{p'}[J_\mu]u_p) =$$

$$= e\frac{1}{4m^2}(\bar{u}_{p'}[2mF_E P_\mu + 4m^2 F_M X_\mu^{(M)}]u_p) \; .$$

Thus we may write $J_\mu a^\mu$ in two equivalent forms

$$J_\mu a^\mu = F_1 \not{a} + \frac{F_2}{2m}\not{q}\not{a} =$$

$$= \frac{F_E}{2m}(Pa) + \frac{F_M}{8m^2}(\not{a}\not{P}\not{q} - \not{q}\not{P}\not{a}) \; . \qquad (2,22)$$

To evaluate the cross-section we have to average $[M_{fi}^+ M_{fi}]$ over the initial spin orientations of the electron and proton and to sum .p the final spins. We decide upon using the original form and in the following draw up the essential steps of this calculation.

$$\frac{1}{2}\sum_{\text{El. spin}} \cdot \frac{1}{2}\sum_{\text{Prot. spin}} [M_{fi}^+ M_{fi}] =$$

$$= \frac{1}{2}\sum_{\text{El}} \frac{(2\pi)^8[\delta^4(p'-p+k'-k)]^2}{16EE'p_o p_o'} \frac{1}{2}\text{tr}\{(\not{p}+m)[F_1\not{a}^+ +$$

$$+ \frac{F_2}{2m}\not{a}^+\not{q}](\not{p}'+m) \times [F_1\not{a} + \frac{F_2}{2m}\not{q}\cdot\not{a}]\} \; . \qquad (2,23)$$

Some "trace rules" following from the general properties of the Dirac-algebra

$$\gamma^\mu\gamma^\nu + \gamma^\nu\gamma^\mu = 2g^{\mu\nu}\cdot 1 \quad \text{with } g^{\mu\nu} = \begin{pmatrix} 1 & & & 0 \\ & -1 & & \\ & & -1 & \\ 0 & & & -1 \end{pmatrix}$$

$$\gamma_0^+ = \gamma_0, \quad \gamma^{k+} = -\gamma^k$$

considerably facilitate the evaluation of the traces.
We have

$$\mathrm{tr}\, 1 = 4$$

$$\mathrm{tr}\,(\gamma^\mu \gamma^\nu) = 4 g^{\mu\nu}$$

$$\mathrm{tr}\,(\gamma^\mu \gamma^\nu \gamma^\rho \gamma^\sigma) = 4\,(g^{\mu\nu} g^{\rho\sigma} - g^{\mu\rho} g^{\nu\sigma} + g^{\mu\sigma} g^{\nu\rho})$$

$$\mathrm{tr}\,(\text{product of an odd number of } \gamma\text{'s}) = 0 \;.$$

Using these formulas in the calculation of the nucleon traces we obtain (apart from a common factor) the following coefficients for the individual combination of the form factors:

$$F_1^2 \propto \tfrac{1}{2} \sum_{El} \{4\,(pa^+)(pa) + q^2 (a^+ a)\}$$

$$\frac{F_1 F_2}{2m} \propto \tfrac{1}{2} \sum_{El} \{4 m q^2 (a^+ a)\}$$

$$\frac{F_2^2}{4m^2} \propto \tfrac{1}{2} \sum_{El} \{4 m^2 q^2 (a^+ a) - 4 q^2 (pa^+)(pa)\} \;.$$

The remaining evaluation of the electron traces yields

$$\tfrac{1}{2} \sum_{El} \{(pa^+)(pa)\} = \tfrac{1}{2}\,\mathrm{tr}\,(\not{k} \not{p} \not{k}' \not{p}) =$$

$$= \left(\frac{e}{q^2}\right)^2 4 E E' p_0^2 \cos^2 \frac{\theta}{2}$$

$$\tfrac{1}{2} \sum_{El} \{a^+ a\} = \tfrac{1}{2}\,\mathrm{tr}\,(\not{k} \gamma^\mu \not{k}' \gamma_\mu) = \left(\frac{e}{q^2}\right)^2 (-E E' \sin^2 \frac{\theta}{2}) \;.$$

We get

$$\frac{1}{4}\sum_{El}\sum_{Prot}[M^+_{fi}M_{fi}] = \frac{(2\pi)^8[\delta^4(p'-p+k'-k)]^2}{16p_op'_o}\frac{e^4}{q^2}\times$$

$$\times\{F_1^2(16p_o^2\cos^2\frac{\theta}{2} - 8q^2\sin^2\frac{\theta}{2}) -$$

$$-\frac{F_1F_2}{2m}32m\sin^2\frac{\theta}{2} + \frac{F_2^2}{4m^2}(-32m^2q^2\sin^2\frac{\theta}{2} -$$

$$-16q^2p_o^2\cos^2\frac{\theta}{2})\} . \tag{2,24}$$

Reducing the square of the δ-function to the δ-function itself we proceed to a finite normalization-volume V for the moment. Now we have to take into account a normalization factor $(V)^{-\frac{1}{2}}$ for each particle-operator.

With the integral representation of the δ-function we obtain

$$[\delta^4(p'-p+k'-k)]^2 = \frac{VT}{(2\pi)^4}\delta^4(p'-p+k'-k) ,$$

in which T represents the reaction duration tending to infinity in a limit.

After averaging over the spins we square the transition amplitude, sum up the permitted phase space and divide it by the reaction duration T and the flux of the ingoing electrons $v_{El}/V = 1/V$.

The resulting expression is equal to the cross-section per unit time i.e.

$$d\sigma = \frac{1}{T}V\cdot\frac{1}{V^4}\frac{\delta(p'-p+k'-k)\cdot VT}{(2\pi)^4}\frac{V}{(2\pi)^3}d^3\vec{p}\,'\times$$

$$\times\frac{V}{(2\pi)^3}d^3\vec{k}'\frac{(2\pi)^8p_o}{p'_o}(\frac{e^2}{q^2})^2\cos^2\frac{\theta}{2}\times$$

$$\times\{(F_1^2 - \frac{q^2}{4m^2}F_2^2) - \frac{q^2}{4m^2}2\tan^2\frac{\theta}{2}(F_1+F_2)^2\} . \tag{2,25}$$

We are interested in the differential cross-section with respect to the electron scattering angle. We integrate over d^3k' and $p_0'^2 dp_0'$

$$\delta^4 (p'-p+k'-k)\, d^3\vec{p}'\, d^3\vec{k}' = E'^2 dE'\, d\Omega_e'\, \delta(E-E' -$$

$$-\sqrt{m^2+\vec{q}^2}+m) \ , \qquad (2,26)$$

in which $\vec{q}^2$ is represented by the expression (2,2).

Integrating further over E' we also use $\delta(f(E)) = \frac{\delta(E'-E_n)}{|f'(E_n)|}$

where E_n is the simple zero of $f(E')$ and get

$$\frac{d}{dE'}(E'-E+\sqrt{m^2+(E-E')^2+4EE'\sin^2\theta/2}-m) =$$

$$= \frac{m}{p_0'}\left(1+\frac{2E}{m}\sin^2\frac{\theta}{2}\right) \ . \qquad (2,27)$$

The use of (2,25), (2,26), (2,27) and (2,1) determines the cross-section as follows

$$\frac{d\sigma}{d\Omega_e'} = \sigma_{Mott}\left(1+\frac{2E}{m}\sin^2\frac{\theta}{2}\right)^{-1}[F_1^2(q^2)+\tau F_2^2(q^2)+$$

$$+2\tau\cdot\tan^2\frac{\theta}{2}(F_1(q^2)+F_2(q^2))^2] \ , \qquad (2,28)$$

with

$$\sigma_{Mott} = \frac{e^4}{(4\pi)^2}\,\frac{1}{4E^2}\,\frac{\cos^2\frac{\theta}{2}}{\sin^4\frac{\theta}{2}} \ . \qquad (2,29)$$

Our result is the well-known Rosenbluth formula, the factors in the brackets of (2,28) clearly demonstrate the corrections. These corrections modify the cross-section for the scattering of ultra-relativistic electrons on an external potential $A_\mu^{ex}(x) = \frac{e}{|\vec{x}|}$ (Mott-scattering) by taking into account the spin of the target and the finite distribution of the charge and the magnetic moments.

Also in contrast to the Mott-scattering on a static external field the Rosenbluth formula contains an additional factor $(1-\frac{2E}{m}\sin^2\frac{\theta}{2})^{-1}$ which corresponds to the recoil of the target. In order to determine the form factors we insert into the left hand side of (2,28) the measured cross-section. Note that the corresponding radiative corrections have to be included.

To determine F_1 and F_2 for a special value of their argument we have to solve the quadratic equation (2,28). Thus we need at least two measurements for the same q^2 but for various initial energies and scattering angles. The so - called "ellipse-method" provides for a suitable tool to handle this problem. We plot each measured cross-section into a $F_1 - F_2$ plane. Several independent measurements yield various ellipses which should intersect in two points. One of these two points is used to determine F_1 and F_2. The inevitable experimental inaccuracies often cause rather large uncertainties in the form factors.

Thus it was proposed [5] to reduce equation (2,28) to principal axis i. e. to get rid of the mixed terms by means of a transformation. The suitable transformation is given by (2,12).

Now we obtain

$$\frac{d\sigma}{d\Omega'_e} = \sigma_{Mott}\,(1+\frac{2E}{m}\sin^2\frac{\theta}{2})^{-1}\left[\frac{G_E^2+\tau G_M^2}{1+\tau} + 2\tau\tan^2\frac{\theta}{2}\cdot G_M^2\right]. \qquad (2,30)$$

The corresponding graphic determination of G_E^2 and G_M^2 is now effected by means of the "line-method" (Schopper), for the relation (2,28) describes a line in a coordinate system with the abscissa $\tan^2\frac{\theta}{2}$ and the ordinate

$$\frac{\frac{d\sigma}{d\Omega'_e}\,(1+\frac{2E}{m}\sin^2\frac{\theta}{2})}{\sigma_{Mott}}\,.$$

G_E^2 represents the value of the ordinate for $\tan^2 \frac{\theta}{2} = -\frac{1}{2+2\tau}$ multiplied by $(1+\tau)$.

The gradient of the line proves to be the value of $2\tau\, G_M^2$. The fact that the Rosenbluth cross-section (2,28) may be transformed into the form (2,30) is a feature common to all processes involving one photon exchange [6]. The form factors defined herein represent the "physical" form factors also from the theoretical point of view.

3. Electromagnetic Form Factors of the Proton

In this section we shall give a brief account of the experimental situation with respect to the form factors of the proton as of summer 1967. An overall picture of the existing data is represented by Figs. 3 and 4, taken from Albrecht [7].

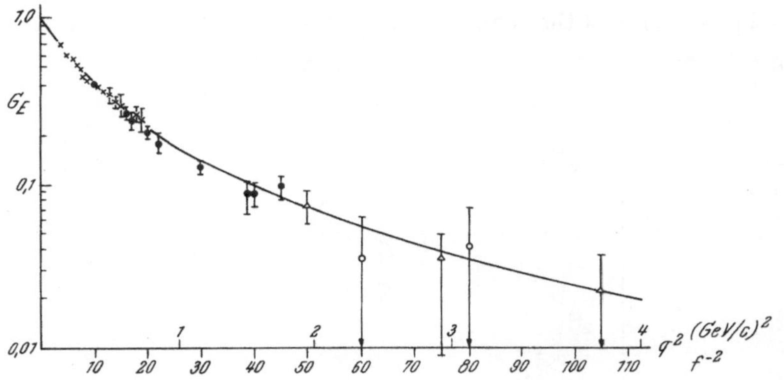

Fig. 3. Electric form factors versus q^2

Fig. 3 shows the electric form factor $G_E(q^2)$ up to values of q^2 of about 4 $(GeV/c)^2$ (or about 102 f^{-2}). For large q^2 the term with G_E in (2,30) is very small and can be detected only for small angles. On the other hand the cross-section is insensitive to large variations of G_E for large q^2 and θ. In this case G_E may even be zero.

In Fig. 4 we have a plot of $G_M(q^2)/\mu_p$, $\mu_p = 2.79$ being the value of the proton's total magnetic moment in Bohr magnetons. Here the

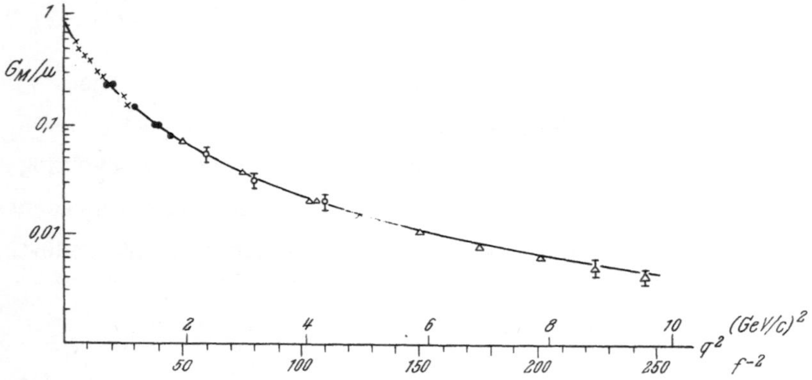

Fig. 4. Magnetic form factor versus q^2

q^2 range extends to about 10 $(GeV/c)^2$ (or about 256 f^{-2}). The solid lines in both cases represent a simple dipole fitted to the data (see also end of this section).

The behavior of the form factors in the region of lower momentum transfer is shown in Fig. 5 (Janssens [8])

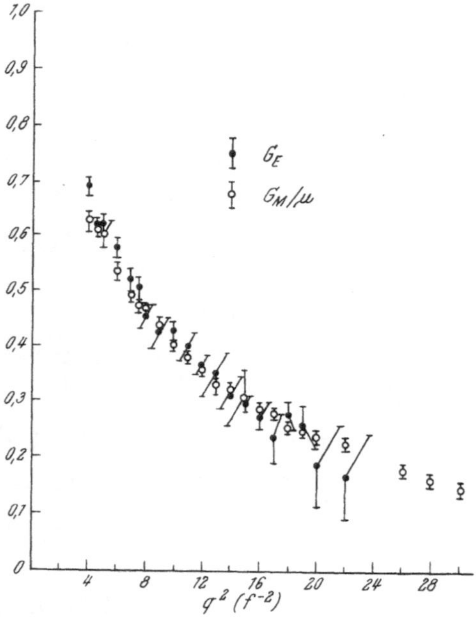

Fig. 5 .The experimental values for the charge and magnetic form factors of the proton as function of q^2

Within the experimental errors the electric and magnetic form factors of the proton (both normalized to unity at $q^2 = 0$) seem to have the same dependence on q^2.

As already mentioned in connection with the introduction of the Dirac and Pauli form factors $F_{1,2}$ (see section II.2) the latter are closely related to the spatial distribution of the electric charge and the anomalous magnetic moment within the proton. Especially in a Lorentz system where the time component of the momentum transfer q_μ vanishes (barycentric system) these form factors are indeed the Fourier transforms of the respective densities. It is, however, not quite clear whether there exists a definite relation between the densities and the real physical extension of the proton.

Bearing this restriction in mind one can define a mean square radius of the spatial distribution by (in the following we are working in the barycentric system unless indicated otherwise):

$$<r^2>_{1,2} = \left(\int r^2 F_{1,2}(\vec{x})\, d^3x \right) / \left(\int F_{1,2}(\vec{x})\, d^3x \right) , \qquad (2,31)$$

where (compare (2,9))

$$F_{1,2}(\vec{x}) = \int \frac{d^3q}{(2\pi)^3}\, e^{-i\vec{q}\cdot\vec{x}}\, F_{1,2}(-\vec{q}^{\,2}) \qquad (2,32)$$

represents the above discussed spatial distributions (or densities) of charge and magnetic moment. We now want to express (2,31) in terms of the form factors $F_{1,2}(-\vec{q}^{\,2})$. First we see from the inverse of (2,32)

$$\int F_{1,2}(\vec{x})\, d^3x = F_{1,2}(-\vec{q}^{\,2}) \Big|_{\vec{q}^{\,2}=0} = F_{1,2}(0) .$$

By choosing an appropriate coordinate system for the integration ($\vec{q}\cdot\vec{x} = |\vec{q}|\, r \cos\theta$) and expanding the exponential for small $|\vec{q}|$ we get, assuming spherical symmetry,

$$F_{1,2}(-\vec{q}^{\,2}) \simeq \int r^2 dr d\varphi d(\cos\theta)\,[1 + i\,|\vec{q}|\,r\cos\theta -$$

$$- \tfrac{1}{2}\,|\vec{q}|^2 r^2 \cos^2\theta + \ldots]\,F_{1,2}(r)\ .$$

Due to the symmetric integration over $\cos\theta$ only terms of even power remain in the bracket; taking the derivative with respect to $(-\vec{q}^{\,2})$ and putting then $\vec{q}^{\,2} = 0$ we get

$$\left.\frac{dF_{1,2}(-\vec{q}^{\,2})}{d(-\vec{q}^{\,2})}\right|_{\vec{q}^{\,2}=0} = \tfrac{1}{2}\int r^2 dr d\varphi d(\cos\theta) r^2 \cos^2\theta\, F(r) =$$

$$= \frac{1}{6}\int r^2 dr d\theta\,(2r^2)\,F(r) = \frac{1}{6}\int r^2 F(\vec{x}) d^3x\ .$$

Therefore (2,31) in terms of the form factors becomes

$$\langle r^2\rangle_{1,2} = 6\,\frac{F'_{1,2}(-\vec{q}^{\,2})\,|\,\vec{q}^{\,2}=0}{F_{1,2}(0)}\ . \tag{2,31a}$$

Here the prime denotes the derivative with respect to the argument as defined above. From (2,31a) it follows that for small $\vec{q}^{\,2}$ we get in our special coordinate system

$$F_{1,2}(-\vec{q}^{\,2}) = F_{1,2}(0)\,[\,1 - \frac{\vec{q}^{\,2}}{6}\langle r^2\rangle_{1,2} + \ldots\,]\ . \tag{2,33}$$

This can be generalized to covariant form which then is valid in any coordinate system:

$$F_{1,2}(q^2) = F_{1,2}(0)\,[\,1 + \frac{1}{6}\,q^2\,\langle r^2\rangle_{1,2} + \ldots\,]\ . \tag{2,34}$$

Finally we want to discuss some of the theoretical attempts to fit the experimental data. One of these, the so-called "dipole fit", may be written as

$$G_E = \frac{G_M}{\mu} = \left(1 + \frac{q^2}{18.1}\right)^{-2} , \qquad (2,35)$$

where q^2 is measured in f^{-2}; it is represented by the solid lines in Figs. 3 and 4. As can be seen, this fit follows very closely the experimental points; its physical interpretation, however, is not obvious. For theoretical reasons it is convenient to introduce linear combinations of the various form factors of the proton and neutron, the so-called isotopic form factors:

$$G_{ES} = \tfrac{1}{2}\left(G_{Ep} + G_{En}\right) ,$$

$$G_{EV} = \tfrac{1}{2}\left(G_{Ep} - G_{En}\right) ;$$

$$G_{MS} = \tfrac{1}{2}\left(G_{Mp} + G_{Mn}\right) ,$$

$$G_{MV} = \tfrac{1}{2}\left(G_{Mp} - G_{Mn}\right) . \qquad (2,36)$$

Here the indices p, n refer to proton and neutron and S, V stand for the isoscalar and isovector parts of the form factor respectively. If one assumes that the isoscalar form factors are dominated by intermediate states coupled to the ω and φ mesons whereas while the isovector part is connected with the ρ meson, these form factors can be written in form of a "3-pole fit" (q^2 measured in f^{-2}):

$$G_{ES} = 0.5\left\{ \frac{S_{E1}}{1+q^2/15.7} + \frac{S_{E2}}{1+q^2/26.7} + (1-S_{E1} - S_{E2})\right\} ,$$

$$G_{MS} = 0.44\left\{ \frac{S_{M1}}{1+q^2/15.7} + \frac{S_{M2}}{1+q^2/26.7} + (1-S_{M1} - S_{M2})\right\} ;$$

$$G_{EV} = 0.5\left\{ \frac{V_{E1}}{1+q^2/M_\rho^2} + (1-V_{E1})\right\} ,$$

$$G_{MV} = 2.353 \{ \frac{V_{M1}}{1+q^2/M_\rho^2} + (1-V_{M1}) \} . \tag{2,37}$$

Here the constant terms represent the contributions from non-resonant states or states of higher mass, and the parameters S_{E1}, S_{E2}, S_{M1}, S_{M2}, V_{E1}, V_{M1} are related to the various coupling strengths. It may be noted that the ω and φ mesons are assigned their well-defined observed masses (783 MeV $\hat{=}$ 3.96 f^{-1} and 1020 MeV $\hat{=}$ 5.16 f^{-1} respectively), whereas usually the ρ meson's mass is treated as an adjustable parameter due to the large observed width of this resonance. In the following table we shall list some results of different authors.

Author	S_{E1}	$-S_{E2}$	S_{M1}	$-S_{M2}$	V_{E1}	V_{M1}	$M_\rho^2(f^{-2})$
Hughes [9]	2.18	1.11	2.42	1.35	1.05	1.05	7.51
Janssens [8]	2.50	1.60	3.33	2.77	1.16	1.11	8.19
Wilson [10]	2.86	1,86	2.41	1.34	1.05	1.04	6.40

Note that the average value observed for the mass of the ρ meson squared is $14.5 f^{-2}$, in contrast to the fitted values above. For a more detailed discussion of the theoretical interpretation of the experimental data see e.g. Wilson [10].

Literature

[1] W.J. Karzas, W.K.R. Watson, F. Zachariasen, Phys.Rev. 110, 253 (1958).

[2] S.D. Drell, F. Zachariasen, Electromagnetic Structure of Nucleons, Oxford: University Press (1961).

[3] R.G. Sachs, Phys.Rev. 126, 1365 (1962).

[4] F.J. Ernst, R.G. Sachs, K.C. Wali, Phys.Rev. 119, 1105 (1960).

[5] H. Schopper, Nuovo Cim. 24, 761 (1962).

[6] M. Gourdin, Nuovo Cim. 21, 1094 (1961).

[7] W. Albrecht, H.J. Behrend, H. Corner, We. Flanger, H. Hultschig, Phys.Rev.Lett. 18, 1014 (1967).

[8] T. Janssens, R. Hofstadter, E.B. Hughes, M.R. Yearian, Phys.Rev. 142, 922 (1966).

[9] E.B. Hughes, T.A. Griffy, M.R. Yearian, R. Hofstadter, Phys.Rev. 139, B458 (1965).

[10] R. Wilson, Proc. of the Int. Symposium on Electron and Photon Interactions at High Energies, Hamburg (1965), Vol. I, p. 43.

III. Determination of the Neutron Form Factors Derived from Quasielastic Electron-Deuteron-Scattering

1. Introduction

In describing the scattering of ultrarelativistic electrons on free neutrons we may of course again apply the Rosenbluth formula (2,28, 2,30). We merely have to replace the proton form factors by the corresponding form factors for the charge- and moment-distribution of the neutron for we should expect the neutron also to be spatially extended. (Approximately a meson Compton-wavelength.) This extension is associated with the finite size of a cloud of virtual mesons centered around the neutron. As will be described later, this interpretation leads to a curious value for the mean charge radius of the neutron when considering the low-energy scattering of neutrons on the electrons surrounding a nucleus. But we find at least a deviation from the point structure with respect to the magnetic moment. This may be shown by comparing the e-n interaction in the electron disintegration of the deuteron with the corresponding e-p scattering [1]. If we assume a point-neutron and take into account the structure of the proton the theory will be incompatible with the experiment. We only get rid of this discrepancy by choosing a suitable magnetic form factor for the neutron. In section II. 3 we have demonstrated that the mean squared radii for the charge and the magnetic moment arise from the gradient of the form factors in the limit $q^2 \to 0$. In case of a proton we obtain these quantities by extrapolating the electron cross-section. Thus the values are not particularly accurate.

With view to the charge distribution radius of the neutron the situation is more favorable. Here very accurate measurements for the scattering of thermic neutrons exist (for a short discussion see Drell [2], p. 20). The interpretation of these experiments yields the repeatedly confirmed value $<r^2>_{1,N} \simeq 0$. This value may be considered the most accurate information on the properties of a neutron form factor we have obtained so far.

The thermic neutron scattering will certainly give no information as to interactions, in other words magnetic interaction.

Our interest in the magnetic structure or the "internal" charge structure ($<r^2>_{1,N}$ describes only the extension of the charge cloud) again leads us to the electron-scattering experiments. But now essential experimental difficulties arise. We have no means to perform a pure e-n scattering experiment because we cannot produce a target consisting of free neutrons. Thus we are forced to scatter the electrons by light nuclei, particularly by deuterons. But now we have to separate the e-n scattering from the total process by means of some subtraction mechanism.
The most successful method of this type up to now has been the inelastic e-d-scattering, that is to say the electrodisintegration of the deuteron:

$$e + d \rightarrow e + p + n \quad .$$

The corresponding Feynman graph

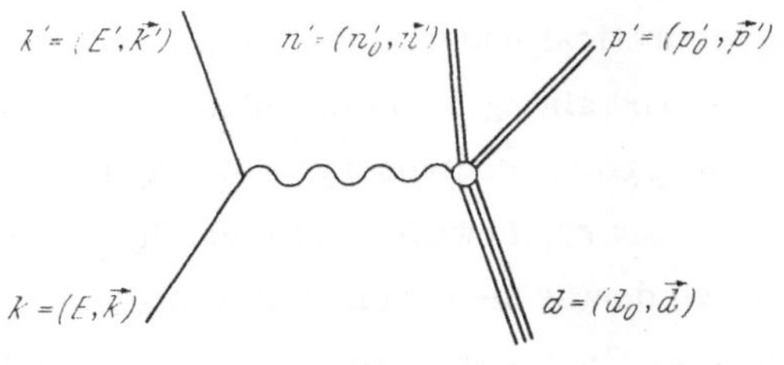

Fig. 6

gives the only contribution, apart from radiative corrections, as long as the electron energy is below the threshold of the electro-meson-production.

We know that the deuteron is the nucleus with the smallest binding energy per nucleon (B = 2.2 MeV). We have every reason to believe that any binding-effects are of minor importance in this case than when scattering electrons by heavy nuclei. Moreover, our process holds great advantage for the experimentator with view to measurements. We are up against certain difficulties in similar cases of electrodisintegration in heavy nuclei in order to distinguish the two-particle decay from three- or multiparticle states of the nucleon system.

Another possibility to gain better knowledge concerning neutron form factors consists in elastic e-d scattering. The theory of the process is essentially based on the deuteron model already used. While on the one hand the inevitable relativistic corrections in the applied electron-energies may be reduced to small (but important) contributions by means of an adroitly chosen kinematics, on the other hand the elastic scattering depends on a relativistic description of the deuteron. In spite of a number of very fine attempts to establish a relativistic theory of the deuteron (Gourdin [3], Gross [4], [5]) the theoretical uncertainties are still greater than in the case of the electrodisintegration in which at least the unbound final state of the proton-neutron-system may be described precisely. It is necessary, however, to investigate also the influence of the respective deuteron-models and to make corrections concerning a possible final state interaction of the nucleons. For this reason we shall confine ourself in the following to the calculation of inelastic e-d scattering.

2. Quasielastic Electron-Deuteron-Scattering

We are mainly interested in the electromagnetic form factors of free nucleons. But we are indeed obliged to rely on the scattering of electrons by weakly bound neutrons in the deuteron. As we are considering a three-particle state the final energy E' of the scattered electron cannot be determined for a fixed electron scattering angle. Thus there exists a certain region in the energy spectrum of the scattered electrons where - in the sense of an "impact approximation" - the definite nucleon interacting with the electron may be regarded as free in first approximation. Here we are dealing with a purely two-particle process in which the non-interacting nucleon plays the part of a "spectator"; the somewhat rough description of this situation explains the notation of "quasielastic scattering". As to the calculated cross-section we shall see later that the electron spectrum possesses a maximum at this very point (apart from a relativistic correction). Here the spatial momentum of the spectator nucleon virtually vanishes in the lab. system. Thus this part of the cross-section $d^2\sigma/dE'd\Omega'_e$ is called "quasielastic peak". The experimentally found number of scattered electrons when plotted against E', the energy of the scattered electron for a fixed lab.-scattering angle, shows besides the broad inelastic ridge also the purely elastic scattered electrons. Under certain circumstances this contribution should be infinitely large at the point

$$E' = E\left(1 + \frac{2E}{M}\sin^2\frac{\theta}{2}\right)^{-1} \qquad M = \text{mass of the deuteron} \qquad (3,1)$$

But it is broadened to a curve of Gaussean shape on account of the uncertainties of the energy in the primary beam.

Because of the bremsstrahlung energy-losses the peak is also distinctly broadened toward the smaller values of E'. This branch of the elastic peak is extending into the region of inelastic scattering and has to be subtracted from the inelastic spectrum in addition to the radiative corrections of the actual electro-disintegration. We may calculate the threshold from the energy conservation and the demand that after the decay both nucleons are at rest in their center of mass system. Thus we define the relative momentum of the nucleons in the lab. system:

$$\vec{p} = \tfrac{1}{2}(\vec{p}\,' - \vec{n}\,') \quad . \tag{3,2}$$

With

$$\vec{q} = \vec{p}\,' + \vec{n}\,' \tag{3,3}$$

the energy-conservation yields:

$$\sqrt{m^2 + \vec{p}\,'^2} + \sqrt{m^2 + \vec{n}\,'} = E - E' + 2m - B \;;$$

to the power m^{-2} we obtain

$$m\left(\frac{1}{2m^2}\left(\frac{\vec{q}}{2} + \vec{p}\right)^2 + \frac{1}{2m^2}\left(\frac{\vec{q}}{2} - \vec{p}\right)^2 + 0\left(m^{-4}\right)\right) \approx \Delta E - B$$

with

$$-\Delta E = E' - E \quad . \tag{3,4}$$

The exact calculation by means of the relativistic kinematics yields (see appendix) to order m^{-2}:

$$\Delta E = \frac{\vec{q}^{\,2}}{4m} + \frac{\vec{p}_c^{\,2}}{m} + B \quad . \tag{3,5}$$

With the requirement $\vec{p} = 0$ we find with (2,2) for the threshold of the inelastic scattering, apart from terms of higher powers than m^{-2},

$$E' = (E-B)\left[1 + \frac{E}{m}\sin^2\frac{\theta}{2}\right]^{-1} \quad . \tag{3, 6}$$

Thus this energy threshold is separated from the elastic peak by an amount somewhat smaller than the binding-energy B. As we have seen the quasielastic peak is given apart from relativistic corrections by $\vec{p}\,' = 0$ or $\vec{n}\,' = 0$, that means it is represented in any case by $|\vec{p}| = |\vec{q}|/2$. With (2,1) we obtain the following expression in order m^{-2}:

$$E'_{max} = (E - B)\left[1 + \frac{2E}{m}\sin^2\frac{\theta}{2}\right]^{-1} . \qquad (3,7)$$

An example of the complete electron spectrum for a definite scattering angle is shown in Fig. 7.

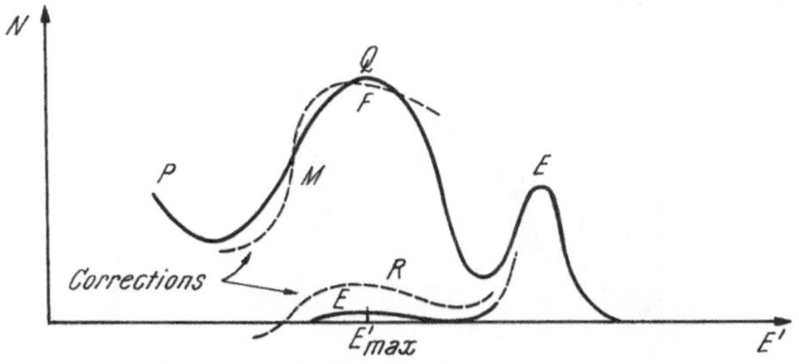

Fig. 7

The ordinate represents the number of the events in arbitrary units. We observe that according to the effects discussed above the elastic peak no longer is completely separated from the broad inelastic peak. Subtracting the radiative corrections (R) and the contribution of the elastic scattering (E) from the total spectrum above we obtain the part due to purely inelastic e-d scattering. This part, however, contains some undesirable contributions. Whereas the increase of the curve in the region (P) indicates the beginning of the electron-pion-production (thus this part of the spectrum cannot be used for practical calculations) there are some secondary processes, namely the scattering by mesonic exchange-currents (M) which appear along the whole spectrum. But it is very difficult to evaluate such effects. Two typical processes

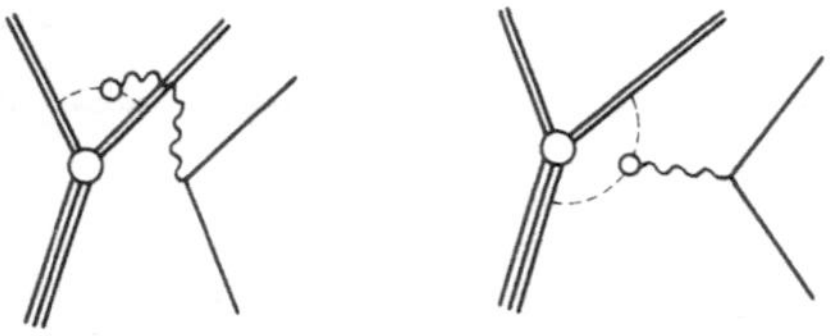

Fig. 8

show clearly that here another scattering mechanism appears which significantly differs from the scattering by a virtual meson effecting only the same nucleon line, as for instance in the case of a quasielastic scattering.

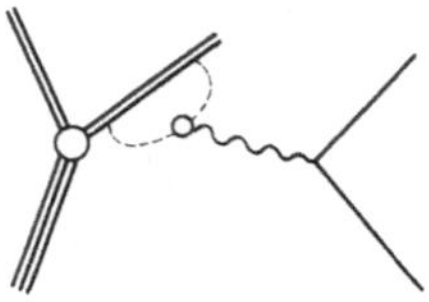

Fig. 9

Indeed, a process of the latter type was taken into account from the beginning by the introduction of form factors. These processes constitute the electromagnetic structure of the nucleon and are summed up symbolically by means of a hatched circle in the diagrams. But the contributions of mesonic exchange-currents obviously defy a consideration in a similar phenomenological way. The present status of the theory of strong interactions does not permit the explicit evaluation of these contributions. Thus we can make some statements about their contributions to the cross-section only by means of rather complicated dispersion-theoretical considerations [6]. Very soon, however, measurements of the double-differential cross-section $d^2\sigma/dE'd\Omega'_e$ at the quasielastic peak led to the opinion that mesonic contributions are minimal there [1].
An additional unsolved problem results from the fact that in the initial state the nucleons are bound in the deuteron and therefore do not constitute free targets. The crucial question, namely to

which extent the resulting form factors for nucleons off the mass shell differ from the form factors for free nucleons, has not been answered up to now. In this case it is again of advantage that we are dealing with a weakly bound system, i.e. the deuteron, in which the nucleons are close to their respective mass shells.

So far we have confined ourselves merely to secondary processes which may be subtracted from the experimental values ((E) and (R)) or which cannot be considered in the following calculations of the cross-section. We therefore shall not take these processes into account. As we shall see in section III.4 the remaining corrections which are responsible for a possible final state interaction of the nucleons may be easily included in this theory. Here, however, we lose the relativistic invariance of the theory (for a detailed discussion see McGee [7]). This last point is the reason why we do not use for our calculations the elegant but complicated covariant formulations of the inelastic e-d scattering [7], [8]. We shall include most of the important consequences of this theory in our semirelativistic calculation at the appropriate point.

3. Calculation of the Cross-Section in the Lab.-System

We are mainly interested in the cross-section differential with respect to the scattering angle and the final energy of the electron: $d^2\sigma/d\Omega'_e dE'$. In order to obtain it we have to integrate the totally differential cross-section over the variables of the nucleons. The starting point of this phase space integration is the transition probability per unit time and flux of the incoming particles:

$$d\sigma = \frac{V}{|v_e|}\,\frac{1}{V^5}\,[(2\pi)^8\delta^4\,(d+k-p'-n'-k')]^2\,\frac{1}{T}\;\cdot$$

$$\cdot\;\frac{V}{(2\pi)^3}\,d^3\vec{k}'\cdot\frac{V}{(2\pi)^3}\,d^3\vec{p}'\,\frac{V}{(2\pi)^3}\,d^3\vec{n}'\;\sum_{spin}|M^d_{fi}|^2 \qquad (3,8)$$

The first factor in (3,8) corresponds to the reciprocal of the flux due to the incoming electrons (with $v_e = c = 1$), the second factor represents the normalization of the particle-operators. For the integration over the nucleon momenta we conveniently introduce the respective relative momenta. From (3,2) and (3,3) we obtain

$$\vec{p}\,' = \tfrac{1}{2}\vec{q} + \vec{p} \tag{3,9}$$

$$\vec{n}\,' = \tfrac{1}{2}\vec{q} - \vec{p} \ . \tag{3,10}$$

Since the Jacobian of the corresponding transformation of variables equals one we get

$$d^3\vec{p}\,'\ d^3\vec{n}\,' = d^3\vec{p}\ d^3\vec{q} \ . \tag{3,11}$$

After splitting up the four-dimensional δ-function into a three-dimensional δ-function representing momentum conservation times a one-dimensional energy conserving δ-function we may integrate over $d^3\vec{q}$ immediately. Note that we now have to insert the value of $\vec{q}$ resulting from momentum-conservation.

Thus we are left with the integration over $\vec{p}$, with $d^3\vec{p} = \vec{p}^{\,2} d\,|\vec{p}\,|\,d\Omega_p$. With the help of the energy δ-function the integration over the absolute value of $\vec{p}$ may be performed. To that end we define a function $f\,([\vec{p}\,])$ by

$$\delta\,(2m - B + E - \sqrt{m^2 + \tfrac{1}{4}\vec{q}^{\,2} + \vec{p}^{\,2} + |\vec{p}\,|\cdot|\vec{q}\,|\cos\theta} -$$

$$-\sqrt{m^2 + \tfrac{1}{4}\vec{q}^{\,2} + \vec{p}^{\,2} - |\vec{q}\,|\ |\vec{p}\,|\cos\theta} -$$

$$- E') \equiv \delta\,[f\,(\,|\vec{p}\,|\,)\,] \ . \tag{3,12}$$

The angle θ as well as the other angular variables entering the calculation are shown in Fig. 10 .

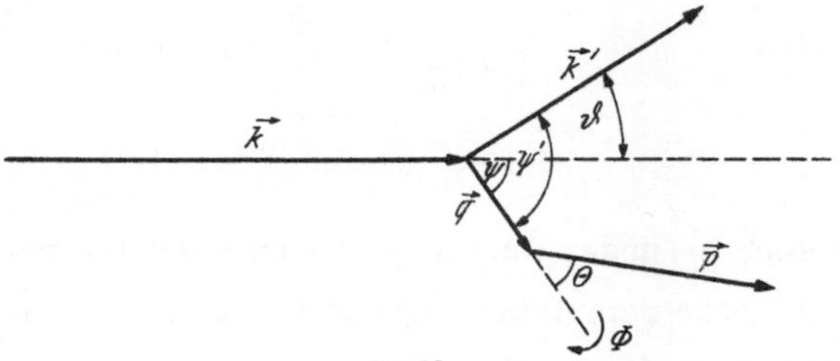

Fig. 10

If we denote the two roots in (3,12) by E'_p and E'_n we obtain with

$$\Delta\epsilon = \Delta E - B \tag{3,13}$$

and

$$E'_n - E'_p = \frac{2|\vec{p}|\ |\vec{q}| \cos\theta}{2m + \Delta\epsilon}$$

the following expression:

$$\left.\frac{1}{f'(|\vec{p}|)}\right|_{f=0} =$$

$$= \frac{E'_p E'_n}{2m|\vec{p}|} \quad \frac{1 + \frac{\Delta\epsilon}{2m}}{(1 + \frac{\Delta\epsilon}{2m})^2 + \frac{\vec{q}^2}{4m^2} \cos^2\theta} \quad . \tag{3,14}$$

Putting $z = \cos\theta$ we get for the double differential cross section

$$\frac{d^2\sigma}{d\Omega'_e dE'} =$$

$$= \frac{1}{(2\pi)^5} \frac{E'_p E'_n}{2m} E'^2 \int_{-1}^{+1} dz \frac{(1+\frac{\Delta\epsilon}{2m}) |\vec{p}|}{(1+\frac{\Delta\epsilon}{2m})^2 + \frac{\vec{q}^2}{4m^2} z^2} \frac{1}{4} \sum_{spins} |M^d_{fi}|^2 \quad .$$

(3,15)

In (3,15) we kept $|\vec{p}|$ under the integral sign since the relevant quantity fixed by the remaining degress of freedom is $|\vec{p}_s|$ and not $|\vec{p}|$. We indeed have (appendix)

$$\vec{p}_s^2 = m \, \Delta\epsilon - \frac{1}{4} \vec{q}^2 \tag{3,16}$$

and

$$|\vec{p}| = |\vec{p}_s| \sqrt{(m^2 + \vec{p}_s + \tfrac{1}{4}\vec{q}^2) / (m^2 + \vec{p}_s^2 + \tfrac{1}{4}\vec{q}^2 \sin^2\theta)} \quad . \tag{3,17}$$

We now insert for the matrix element in question the relativistic e-n matrix elements where we take the respective nucleon momentum in the deuteron as the relative momenta of the nucleons in the initial state. By taking the average over the momentum distribution of the nucleons in the deuteron these e-n matrix elements have to be multiplied by the corresponding probability weights, i.e. by the respective wave functions of the deuteron. Since the influence of the deuteron D-state essentially consists in a reduction of the total-cross-section by the factor $(1-P_D)$ [9], [10] it is sufficient to use pure S-state wave functions for the deuteron

$$\psi_d(r) = \frac{1}{\sqrt{4\pi}} \frac{u(r)}{r} \tag{3,18}$$

with the normalization

$$\int u^2(r)\, dr = 1-P_D \quad . \tag{3,19}$$

In momentum space we obtain

$$|M_{fi}^{(d)}|^2 =$$

$$= |<\vec{p}', \vec{n}' | M^d | \vec{d}>|^2 =$$

$$= \{ <\vec{p}' | M^{(p)} | \vec{p}'-\vec{q}> \tilde{u}(\vec{p}'-\vec{q}) +$$

$$+ <\vec{n}' | M^{(n)} | \vec{n}'-\vec{q}> \tilde{u}(\vec{n}'-\vec{q}) \}^2 =$$

$$= | <\vec{p} + \tfrac{1}{2}\vec{q} | M^{(p)} | \vec{p} - \tfrac{1}{2}\vec{q}> \tilde{u}(\vec{p}-\tfrac{1}{2}\vec{q}) +$$

$$+ <-\vec{p}+\tfrac{1}{2}\vec{q} | M^{(n)} | -\vec{p}-\tfrac{1}{2}\vec{q}> \tilde{u}(-\vec{p}-\frac{\vec{q}}{2}) \}^2 =$$

$$= \tilde{u}^2(\vec{p}-\tfrac{1}{2}\vec{q}) \, |<\vec{p}+\tfrac{1}{2}\vec{q} | M^{(p)} | \vec{p}-\tfrac{1}{2}\vec{q}>|^2 +$$

$$+ \tilde{u}^2(\vec{p}+\tfrac{1}{2}\vec{q}) \, |<-\vec{p}+\tfrac{1}{2}\vec{q} | M^{(n)} | -\vec{p}-\tfrac{1}{2}\vec{q}>|^2 +$$

$$+ \tilde{u}(\vec{p}-\tfrac{1}{2}\vec{q}) \, \tilde{u}(\vec{p}+\tfrac{1}{2}\vec{q}) \, (<\vec{p}+\tfrac{1}{2}\vec{q} | M^{(p)} | \vec{p}-\tfrac{1}{2}\vec{q}>^* \cdot$$

$$\cdot <-\vec{p}-\tfrac{1}{2}\vec{q} | M^{(n)} | -\vec{p}+\tfrac{1}{2}\vec{q}> +$$

$$+ <-\vec{p}-\tfrac{1}{2}\vec{q} | M^{(n)} | -\vec{p}+\tfrac{1}{2}\vec{q}>^* <\vec{p}+\tfrac{1}{2}\vec{q} | M^{(p)} | \vec{p}-\tfrac{1}{2}\vec{q}>) . \tag{3,20}$$

Here $\tilde{u}(\vec{k})$ denotes the Fourier-transform of the deuteron wave function ψ_d.

First we calculate the direct terms in a way analogous to the derivation of the Rosenbluth formula. The proton-current operator has the form (see 2,14)

$$J_p^\mu = e F_E^p \frac{1}{2m} P_p^\mu + e F_M^p \frac{1}{8m^2} (\gamma^\mu \not{P}_p \not{q} - \not{q} \not{P}_p \gamma^\mu) . \tag{3,21}$$

With (2,20) we obtain

$$P_{p,o} = \sqrt{m^2 + (\vec{p} - \tfrac{1}{2}\vec{q})^2} + \sqrt{m^2 + (\vec{p} + \tfrac{1}{2}\vec{q})^2}$$

and identify the two square roots with E'_n and E'_p in (3,12). Using energy conservation we also find

$$P_{p,o} = 2m + \Delta\epsilon \ .$$

Equation (2,20) yields $\vec{P}_p = 2\vec{p}$, which means we now have the four vector

$$P_p = (2m + \Delta\epsilon,\ 2\vec{p}) \ . \tag{3,22}$$

The corresponding relations for the term due to the neutron are

$$J_n^\mu = eF_E^n \frac{1}{2m} P_n^\mu + eF_M^n \frac{1}{8m^2} (\gamma^\mu \not{P}_n \not{a} - \not{a} \not{P}_n \gamma^\mu) \tag{3,23}$$

with

$$P_n = (2m + \Delta\epsilon,\ -2\vec{p}) \ . \tag{3,24}$$

In order to obtain the transition probabilities we have to evaluate as in equations (2,22) - (2,24) the matrix elements of $J_\mu a^\mu$ together with a summation over spins. Omitting for the moment the nucleon indices the summation over the nucleon spins leads to the following coefficients multiplying the square of the form factors (mixed products of these do not occur, see (2,30)):

$$\frac{F_E^2}{4m^2} \propto (Pa^+)(Pa) \cdot P^2$$

$$\frac{F_M^2}{16m^4} \propto -\tfrac{1}{2} P^2 q^2 \left[2(Pa^+)(Pa) - P^2 (a^+a) \right] -$$

$$-\tfrac{1}{2} q^4 P^2 (a^+a) + 2m^2 q^2 P^2 (a^+a) \ .$$

The remaining trace operations can be performed easily:

$$\frac{1}{2} \sum_{\text{spin El}} (Pa^{+})(Pa) = 2\{2(Pk)(Pk') - P^2(kk')\} \cdot \left(\frac{e^2}{q^2}\right)^2$$

$$\frac{1}{2} \sum_{\text{spin El}} (a^{+}a) = 2q^2 \left(\frac{e^2}{q^2}\right)^2 .$$

Considering the respective normalization factors $(16EE'E_pE'_p)^{-1}$ and $(16EE'E_nE'_n)^{-1}$ we obtain

$$|M^p_{fi}|^2 = \frac{P_p^2}{4E_pE'_p} \frac{1}{(1+\tau)^2} \{G_E^{p\,2} \cdot A_p + \tau G_M^{p\,2} \cdot B_p\} \frac{e^4}{q^4} \tilde{u}^2 (\vec{p} - \tfrac{1}{2}\vec{q}) \tag{3,25}$$

$$|M^n_{fi}|^2 = \frac{P_n^2}{4E_nE'_n} \frac{1}{(1+\tau)^2} \{G_E^{n\,2} \cdot A_n + \tau G_M^{n\,2} \cdot B_n\} \frac{e^4}{q^4} \tilde{u}^2 (\vec{p} + \tfrac{1}{2}\vec{q}) \tag{3,26}$$

$$A_j = \frac{2(P_jk)(P_jk') - P_j^2(kk')}{8m^2EE'} , \quad j = p, n , \tag{3,27}$$

$$B_j = \frac{2(P_jk)(P_jk') + P_j^2(kk')}{8m^2EE'} , \quad j = p, n . \tag{3,28}$$

The calculation of the interference terms requires the transition to two-component spinors due to the separation of the two spin spaces. Thus we are forced to use nonrelativistic approximations for the nucleon currents $X_{\mu}^{(E)}$ and $X_{\mu}^{(M)}$ in (3,20).

Whereas the original calculations of the inelastic e-d scattering made use of a suitably chosen Hamiltonian from the beginning - even for the direct terms - [11], [12], [13], in this calculation the different, partly disputed aspects of the approximations influence only the interference terms which are negligible compared with the direct terms. The two nonrelativistic approximations mainly used are derived either from the separation into large and small spinor components in the relativistic pole terms [8], [9] or from a Foldy-Wouthuysen transformation of the relativistic currents and interaction Hamiltonians respectively [14], [15], [16]. We shall apply the latter method, especially the Foldy--Interaction picture. For a thorough discussion of this method see chapter I and the original papers of P. Breitenlohner ([17], [18]).

We are now interested in the Foldy-Interaction transformation of the terms

$$\gamma_{0} X_{\mu}^{(E)} a^{\mu} = a_{0} - \frac{1}{2m} \{ (\gamma q) a_{0} + \beta (Pa) \} \tag{3,29}$$

and

$$\begin{aligned} \gamma_{o} X_{\mu}^{(M)} a^{\mu} = {} & \frac{1}{2m} (\gamma q) a_{0} - \frac{i}{2m} (\tau q a) + \\ & + \frac{\vec{q}^{2}}{4m^{2}} (a_{0} - (\alpha a)) + \frac{1}{4m^{2}} \{ p^{2} (\alpha a) - \\ & - (\alpha P)(Pa) - i \delta (P, q, a) \} . \end{aligned} \tag{3,30}$$

To order m^{-2} the transformed operators have the form

$$\gamma_{0} X_{\mu}^{(E)'} a^{\mu} =$$

$$= a_o \{1 + \vec{q}^2 / 8m^2\} + a_o \frac{i}{8m^2} (\vec{\sigma}, \vec{P}, \vec{q}) -$$

$$- \frac{1}{2m} (\vec{P}\vec{a}) \beta \qquad (3,31)$$

and

$$\gamma_o X_\mu^{(M)'} . a^\mu = - \frac{i}{4m^2} (\vec{\sigma}, \vec{P}, \vec{q}) a_o - \frac{i}{2m} (\vec{\tau}, \vec{q}, \vec{a}) . \qquad (3,32)$$

The matrices $\vec{\sigma}$, β, $\vec{\tau}$, δ, $\vec{\alpha}$, $\vec{\gamma}$ appearing in (3,29) - (3,32) may be represented by the well-known Pauli matrices in the following way:

"odd operators" "even operators"

$$\delta = \begin{pmatrix} 0 & 1 \\ 1 & 0 \end{pmatrix} ; \vec{\alpha} = \begin{pmatrix} 0 & \vec{\sigma} \\ \vec{\sigma} & 0 \end{pmatrix} \qquad \vec{\sigma} = \begin{pmatrix} \vec{\sigma} & 0 \\ 0 & \vec{\sigma} \end{pmatrix}, \beta = \begin{pmatrix} 1 & \\ & -1 \end{pmatrix},$$

$$\vec{\gamma} = \begin{pmatrix} 0 & \vec{\sigma} \\ -\vec{\sigma} & 0 \end{pmatrix} \qquad \vec{\tau} = \begin{pmatrix} \vec{\sigma} & 0 \\ 0 & -\vec{\sigma} \end{pmatrix} .$$

According to the prescription of the Foldy-Interaction picture we retained the "even" terms in (3,31) and (3,32) only, due to the fact that we are interested in the left hand upper quadrant of the interaction operator [18].

In order to apply the well-known rules concerning traces in our summation over spins we have to insert the projection operator (to total spin 1) in the p- and n-spin space respectively between the operators of the initial two-nucleon system and obtain

$$\sum_{spins} (M_{fi}^p M_{fi}^{n^+} + M_{fi}^n M_{fi}^{p^+}) = \frac{1}{3} \operatorname{tr} \{ 0^p P_1 0^n \} \qquad (3,33)$$

with

$$0^i = a^\mu X_\mu^{(E)'} F_E^i + a^\mu X_\mu^{(M)'} F_M^i$$

and

$$P_1 = \frac{1}{4}(3 - \vec{\sigma}_p \vec{\sigma}_n) \ .$$

Omitting the normalization factors which later on will be taken over from the direct terms we obtain

$$\sum_{\text{spins}} (M^p_{fi} M^{n^+}_{fi} + M^n_{fi} M^{p^+}_{fi}) = \sum_{\text{spin el}} \frac{1}{(1+\tau)^2} \cdot$$

$$\cdot \frac{1}{3} \mathrm{tr} \{ 2G^p_E G^n_E [(1 + \frac{\tau}{2} + \frac{i}{8m^2}(\vec{\sigma}_p \vec{P}_p \vec{q})) \cdot a_o -$$

$$- (\vec{P}_p \vec{a})/2m] \cdot P_1 \cdot [(1 + \frac{\tau}{2} + \frac{i}{8m^2}(\vec{\sigma}_n \vec{P}_n \vec{q})) a^+_o -$$

$$- (\vec{P}_n \vec{a}^+)/2m] \} + \frac{1}{4m^2} (\vec{\sigma}_p \vec{q} \vec{a}) P_1 (\vec{\sigma}_n \vec{q} \vec{a}^+) \cdot$$

$$\cdot 2 G^p_M G^n_M + 0 (m^{-3}) \ . \tag{3,34}$$

To order m^{-2} the terms with $G_E G_M$ and $(\vec{\sigma}\vec{P}\vec{q})$. $G_E G_E$ vanish because they are connected with an odd number of σ_p and σ_n-matrices leading to spin zero. Two types of traces appear in (3,34). For their evalution we apply the rules

$$\mathrm{tr}\, P_s = (2s + 1)$$

$$\mathrm{tr}\, (\vec{c}_1 \vec{\sigma}_p + \vec{c}_2 \vec{\sigma}_n) P_1 (\vec{c}_3 \vec{\sigma}_p + \vec{c}_4 \vec{\sigma}_n) = (\vec{c}_1 \vec{c}_3 + \vec{c}_2 \vec{c}_4) +$$

$$+ \tfrac{1}{2} (\vec{c}_2 \vec{c}_3 + \vec{c}_1 \vec{c}_4) \ .$$

We then find

$$\frac{1}{3} \mathrm{tr}\, P_1 = 1$$

and

$$\frac{1}{3}\operatorname{tr}(\vec{\sigma}_p, \vec{q}, \vec{a}) P_1 (\vec{\sigma}_n, \vec{q}, \vec{a}) =$$

$$= \frac{1}{3}[\vec{q}^{\,2}(\vec{a}\,\vec{a}^{+}) - q_o^2 (a_o a_o^{+})] .$$

With (3,22) and (3,24) we obtain from (3,34)

$$\sum_{\text{spins}} (M_{fi}^{p} M_{fi}^{n^{+}} + M_{fi}^{n} M_{fi}^{p^{+}}) =$$

$$= \sum_{\text{spin el}} \{ \frac{2 G_E^p G_E^n}{(1+\tau)^2} [(1+\tau)(a_o a_o^{+}) -$$

$$- \frac{(\vec{p}\vec{a}^{+})(\vec{p}\vec{a})}{m^2} + (1+\frac{\tau}{2}) \frac{1}{m}[a_o^{+}(pa) - (pa^{+})a_o]] +$$

$$+ \frac{2 G_M^p G_M^n}{(1+\tau)^2} \frac{1}{3} \frac{1}{4m^2} [\vec{q}^{\,2}(\vec{a}\vec{a}^{+}) - q_o^2 (a_o a_o^{+})] \} .$$

(3,35)

With $a_\mu = (\bar{u}_k, \gamma_\mu u_k)$ and the trace rules (see chapter II) we get after a short calculation the remaining electron traces. Taking into account the normalization factor $(4EE')^{-1}$ we then have

$$\tfrac{1}{2}\operatorname{tr}(a_o a_o^{+}) = \cos^2 \frac{\theta}{2} \frac{e^4}{q^4}$$

$$\tfrac{1}{2}\operatorname{tr}(\vec{a}\vec{a}^{+}) = (\cos^2 \frac{\theta}{2} + 2 \sin^2 \frac{\theta}{2}) \cdot \frac{e^4}{q^4}$$

$$\tfrac{1}{2}\operatorname{tr}(\vec{p}\vec{a})\, a_o^{+} = \frac{|\vec{p}|}{2} \{\cos(kp) + \cos(k'p)\} \frac{e^4}{q^4}$$

$$\tfrac{1}{2}\operatorname{tr}(\vec{a}\vec{p})(\vec{a}^{+}\vec{p}) = \vec{p}^{\,2} \{\cos(kp)\cos(k'p) + \sin^2 \frac{\theta}{2}\} \frac{e^4}{q^4} .$$

In order to express the trigonometric function by means of the scattering parameters of interest a lengthy calculation [19] is needed which leads to the following relations

$$\cos\psi = \frac{\Delta E}{|\vec{q}|}\cos^2\frac{\theta}{2} + \sqrt{1-\{(\Delta E/|\vec{q}|)^2\}\cos^2\frac{\theta}{2}}\,\sin\frac{\theta}{2}\;;$$

$$\cos\psi' = \frac{\Delta E}{|\vec{q}\,'|}\cos^2\frac{\theta}{2} - \sqrt{1-\{(\Delta E/|\vec{q}\,'|)^2\}\cos^2\frac{\theta}{2}}\,\sin\frac{\theta}{2}\;.$$

Averaging over the angle (due to the integration over $d\varphi$ already performed in the phase space integration $d\Omega_p$) we obtain for the combinations of trigonometric functions in question:

$$\cos(pk) + \cos(pk') \rightarrow \cos\Theta\,\frac{2\Delta E}{|\vec{q}|}\cos^2\frac{\theta}{2} \tag{3,36}$$

$$\cos(pk)\cos(pk') \rightarrow \cos^2\Theta\cos\psi\cos\psi' + \tfrac{1}{2}\sin^2\Theta\sin\psi\sin\psi' =$$

$$= \left\{\left(\frac{\Delta E}{|\vec{q}|}\right)^2 + \left[\tfrac{1}{2} - \tfrac{3}{2}\left(\frac{\Delta E}{|\vec{q}|}\right)^2\right]\sin^2\Theta\right\}\cos^2\frac{\theta}{2} + \left[\sin^2\Theta - 1\right]\sin^2\frac{\theta}{2}\;. \tag{3,37}$$

If we write the interference terms in the form

$$2\,G_E^p G_E^n F(z) F(-z) A_i + \tfrac{2}{3} t\, G_M^p G_M^n F(z) F(-z) B_i$$

with

$$F(z) = \tilde{u}\,(\,|\vec{p} - \tfrac{1}{2}\vec{q}\,|\,)\ ;\ F(-z) = \tilde{u}\,(\,|\vec{p} + \tfrac{1}{2}\vec{q}\,|\,)\ , \tag{3,38}$$

the last results yield

$$A_i = (1+\frac{\tau}{2})^2 \cos^2\frac{\theta}{2} - \frac{1}{m^2}\vec{p}^{\,2}\,\{[\,(\frac{\Delta E}{|\vec{q}|})^2 + \\ + [\,\tfrac{1}{2} - \frac{3}{2}(\frac{\Delta E}{|\vec{q}|})^2\,]\sin^2\Theta\,]\cos^2\frac{\theta}{2} + \\ + \sin^2\Theta \sin^2\frac{\theta}{2}\} \tag{3,39}$$

$$B_i = \frac{\tau}{3}\,(\cos^2\frac{\theta}{2} + 2\sin^2\frac{\theta}{2} + 0\,(\tau)) \tag{3,40}$$

$$A_{p,n} = \Big\{[\,1 + (1 - \frac{2\,|\vec{p}|}{|\vec{q}|})\,\frac{\Delta E}{2m}\,]^2 - \\ - \frac{2\,|\vec{p}|}{|\vec{q}|}\,(1 + \frac{\Delta E}{2m})\,(1 \mp \cos\Theta) + \\ + \frac{\vec{p}^{\,2}}{2m^2}\,[\,1 - 3\,(\frac{\Delta E}{|\vec{q}|})^2\,]\sin^2\Theta\Big\}\cos^2\frac{\theta}{2} + \\ + \frac{\vec{p}^{\,2}}{2m^2}\sin^2\Theta \cdot 2\sin^2\frac{\theta}{2} \tag{3,41}$$

$$B_{p,n} = \Big\{[\,1 + (1 - \frac{2\,|\vec{p}|}{|\vec{q}|})\,\frac{\Delta E}{2m}\,]^2 - \\ - \frac{2\,|\vec{p}|}{|\vec{q}|}\,(1 + \frac{\Delta E}{2m})\,(\,1 \mp \cos\Theta\,) + \\ + \frac{p^2}{2m^2}\,[\,1 - 3\,(\frac{\Delta E}{|\vec{q}|})^2\,]\sin^2\Theta\Big\}\cos^2\frac{\theta}{2} +$$

$$+\{(1+\frac{\Delta E}{2m})^2+\frac{\vec{p}^2}{m^2}+\frac{\vec{p}^2}{2m^2}\sin^2\Theta\}2\sin^2\frac{\theta}{2}\,.$$

(3,42)

Thus $A_{p,n}$ differs from A_i by terms of order m^{-3} (apart from the replacement $\sin^2\Theta \to -\sin^2\Theta$), and we may replace the coefficients of A_i by the corresponding terms in $A_{p,n}$ without restriction in accuracy. In $B_{p,n}$ and B_i only terms of zero order contribute to the result to order m^{-2} since we have factored out τ. We may therefore try to put B_i into a form resembling $B_{p,n}$.

We write

$$A_i=\{[1+(1-\frac{2|\vec{p}|}{|\vec{q}|})\frac{\Delta E}{2m}]^2-$$

$$-\frac{\vec{p}^2}{2m^2}[1-3(\frac{\Delta E}{|\vec{q}|})^2]\sin^2\Theta\}\cos^2\frac{\theta}{2}-$$

$$-\frac{\vec{p}^2}{2m^2}\sin^2\Theta\quad\sin^2\frac{\theta}{2} \tag{3,43}$$

$$B_i=\{[1+(1-\frac{2|\vec{p}|}{|\vec{q}|})\frac{\Delta E}{2m}]^2-$$

$$-\frac{\vec{p}^2}{2m^2}[1-3(\frac{\Delta E}{|\vec{q}|})^2]\sin^2\theta\}$$

$$+[(1+\frac{\Delta E}{2m})^2-\frac{\vec{p}^2}{m^2}-\frac{\vec{p}^2}{2m^2}\sin^2\Theta]\,2\sin^2\frac{\theta}{2}\,.$$

(3,44)

Providing A_i and B_i with the corresponding normalization factors $(16\,E_pE'_pE_nE'_n)^{-\frac{1}{2}}(4EE')^{-1}$ and observing that $E_p = E'_n$ and $E_n = E'_p$ we obtain the double differential cross section by combining (3,15), (3,25), (3,26), (3,38) and (3,44) :

$$\frac{d^2\sigma}{dE' d\Omega'_e} = \sigma_{Mott} \frac{m^2 |\vec{p}_s|}{E_s \pi} I(\theta) \tag{3,45}$$

with

$$I(\theta) = \frac{I_1}{1+\tau} + I_2 \, 2 \tan^2 \frac{\theta}{2} \tag{3,46}$$

$$I_i = E_i \left(G_E^{p\,2} + G_E^{n\,2} \right) + \tau M_i \left(G_M^{p\,2} + G_M^{n\,2} \right) +$$

$$+ E_i' \, 2 G_E^p G_E^n + \frac{1}{3} \tau M_i' \, 2 G_M^p G_M^n \tag{3,47}$$

$$E_1 = M_1 = FF1 + \frac{\Delta E}{m} \left(1 + \frac{\Delta E}{2m}\right) FFD + \left(1 - 3 \left(\frac{\Delta E}{|\vec{q}|}\right)^2\right) FFS \tag{3,48}$$

$$E_2 = \frac{FFS}{\tau + 1} \tag{3,49}$$

$$M_2 = \frac{1}{1+\tau} \left[FF2 + FFS \right] \tag{3,50}$$

$$\{FF1, \; FF2, \; FFD, \; FFS\} =$$

$$= \frac{1}{8\pi} \frac{E_s}{m} \int dz \, \frac{C_1 F^2(z) \, |\vec{p}| / |\vec{p}_o| \, (1 + (\Delta E / 2m))}{(1 + (\Delta E / 2m))^2 - (\vec{q}^{\,2} / 4m^2) z^2} \, \cdot$$

$$\cdot \left\{ C_2, C_1, \; \frac{2 |\vec{p}|}{|\vec{q}|} (1 - z), \; \frac{\vec{p}^{\,2}}{2m} (1 - z^2) \right\} \tag{3,51}$$

$$C_1 = \frac{\left(1 + \frac{\Delta E}{2m}\right)^2 - \frac{\vec{p}^{\,2}}{m^2}}{1 + \tau} \tag{3,52}$$

$$C_2 = \left[1 + \left(1 - \frac{2 |\vec{p}|}{|\vec{q}|}\right) \frac{\Delta E}{2m} \right]^2 . \tag{3,53}$$

Replacing

$$\{FF1,\ FF2,\ FFS,\ FFD\} \rightarrow \{FF1',\ FF2',\ -FFS',\ 0\} \qquad (3,54)$$

we may calculate E_i' and M_i' from E_i and M_i. Here FF1', FF2', FFS' are defined by the same expressions as the unprimed quantities (3,51), (3,52) and (3,53) where we have to replace $F^2(z)$ in (3,51) by the mixed product $F(z)F(-z)$. Equation (3,45) - (3,54) represent the result of our semirelativistic calculation of the quasielastic e–d scattering. From its derivation we observe that equation (3,45) will certainly reproduce the correct electron-spectrum even in a neighborhood of the quasielastic peak $|\vec{p}_c| = = \frac{1}{2}|\vec{q}|$.

Some corrections, however, discussed in the following chapter, have to be applied to this formula.

But before doing this we have to consider the functional form of the deuteron wave function and the influence of the deuteron model on the cross-section.

4. Influence of the Deuteron Model

As was shown by Breitenlohner [20] there is practically no difference between cross-sections evaluated with realistic, numerically given deuteron wave functions or analytic Hulthèn functions whose parameters are adjusted to the deuteron parameters of the corresponding realistic functions.

We therefore start from the configuration space function (3,18) with

$$u(r) = \{\ \bar{N} \cdot e^{-\alpha r_c}\left(e^{-\alpha(n-r_c)} - e^{\beta(r-r_c)}\right) \quad \begin{matrix} r < r_c \\ r > r_c \end{matrix} \qquad (3,55)$$

with $\bar{N}$ given by

$$1 - P_D \doteq \bar{N}^2\, e^{-2\alpha r_c}\, \frac{(\beta-\alpha)^2}{2\alpha\beta(\alpha+\beta)} \ ; \qquad (3,56)$$

β and r_c (hardcore) are the above mentioned parameters and $\alpha =$ $= 0.2317$. We find the momentum space wave function by means of a Fourier transformation, where we expand the exponential $e^{i\vec{Q}.\vec{r}}$ into spherical harmonics and apply the orthonormality formulae for the latter. The result is

$$u(|\vec{Q}|) = 2\sqrt{\pi}N\, e^{-\alpha r_c}(\alpha-\beta)F(\pm z), \tag{3,57}$$

where

$$F(\pm z) = \frac{1}{(\alpha^2+Q_\pm^2)(\beta^2+Q_\pm^2)}\{(\beta+\alpha)\cos(r_c Q_\pm) + $$
$$+ (\alpha\beta - Q_\pm^2)\frac{\sin(r_c Q_\pm)}{Q_\pm}\} . \tag{3,58}$$

$Q_\pm$ is given by $Q_\pm = (p^2 + \frac{1}{4}q^2 \mp pq\cos\theta)^{\frac{1}{2}}$. (3,59)

For peak conditions, that is for $p = q/2$, it is necessary to have a series of $F(\theta)$ for $\theta \simeq 0$ to avoid computational difficulties. One finds

$$F(\theta) \sim [\alpha^2\beta^2(\beta-\alpha)]^{-1} \sum_{n=0}^{2}\sum_{l=0}^{2n+1}\left(\frac{Q^2}{\alpha^2\beta^2}\right)^n \frac{(\alpha\beta r_c)^l}{l!}(\beta^{2n+2-l}-\alpha^{2n+2-l}) . \tag{3,60}$$

To find the influence of different deuteron parameters, such as D-state probability, effective range and quadrupole moment, we have adapted the free Hulthèn parameters to the deuteron parameters of various tabulated Kramer-Glendenning wave functions [21]. We have chosen potentials 2, 3, 4, 6, 8 and 9 of these authors and have calculated $E_1(1-P_D)^1$, for various values of $q/2$ and p near the quasielastic peak. In Table 1 we give two different figures for each value of $q/2$ and p. The first (and larger one) gives the maximal relative difference between any two values of $E_1 \cdot (1-P_D)^{-1}$ out of the six deuteron models, the second one gives the maximal difference if we drop potential 3. (The exceptional properties of the latter were also found by Breitenlohner [20] and could be due to its relatively large effective range parameter.)

From Table 1 we may deduce that for reasonable deuteron wave functions the relative uncertainty should never exceed one percent. Thus we proceed to investigate the final state corrections choosing as deuteron wave function that Hulthèn function which corresponds to the Kramer-Glendenning potential 9. Its parameters are: $r_c =$ $= 0.4329 f$, $P_D = 7.425\ \%$, ρ (effective range) $= 1.715 f$.

Table 1. The influence of the D- state probability for various values of q/2 and p near the quasielastic peak

q/2	p	$E_1(1-P_D)^{-1}$	$E_1(1-P_D)^{-1}$ without pot. 3
	1.15	1.1 %	0.5 %
	1.2519	1.7	0.8
1.3449	1.3449	1.7	0.8
	1.4315	1.1	0.5
	1.5300		0.3
	1.3449	1.3 %	0.6 %
	1.4315	1.8	0.8
1.5300	1.53	1.6	0.7
	1.6285	0.7	0.3
	1.7363		
	1.53	1.5 %	0.7 %
	1.6285	1.8	0.9
1.7363	1.7363	1.4	0.7
	1.8697	0.3	0.2
	1.9948	0.3	0.2
	1.7363	1.5 %	0.7 %
	1.8697	1.8	0.8
1.9948	1.9948	0.9	0.4
	2.1262	1.3	0.7
	2.25	4.5	2.6

5. Final State Corrections

To calculate these important corrections to the Born terms of section III.3 we have to formulate the whole process in the framework of the nonrelativistic scattering theory. For this purpose one has to express the electron-nucleon interaction by a proper Hamiltonian and to replace the plane wave two-nucleon final state by the solution of a Schrödinger equation with a nucleon-nucleon potential. The interaction Hamiltonian can be found by a Foldy-Wouthuysen transformation or by performing the nonrelativistic limit of (3,21) and (3,23). In the second case one winds up with the following nonrelativistic currents [9]:

$$J_o^{(i)} = -4m^2 \sqrt{2}\, F(m^2)\, \chi_S^{m'} [F_{1i} - \frac{\vec{q}^{\,2}}{8m^2} (2\varkappa_i F_{2i} - F_{1i})] \chi_1^m \tag{3,61}$$

$$\vec{J}^{(i)} = -2m\sqrt{2}\, F(m^2)\, \chi_S^{m-1} [(F_{1i} + \varkappa_i F_{2i})\, \vec{\sigma}_i \times \vec{q} - i\vec{q} F_{1i}] \chi_1^m , \tag{3,62}$$

with

$$F(m^2) = \sqrt{\frac{8\pi}{m}} (1 + \frac{1}{\sqrt{2}}\omega)\, N \frac{1}{\sqrt{1-\omega^2}} , \tag{3,63}$$

$$N = \sqrt{\frac{2\alpha}{1-\alpha\rho}} \tag{3,64}$$

and $\omega \simeq 0.03$. We have introduced the anomalous magnetic moment $\varkappa_i$ of the i-th nucleon.

Putting these operators between the deuteron state at the right and the solution of the two-nucleon Schrödinger equation at the left one can calculate the transition amplitude for the disintegration process with final state corrections.

Before taking the square of this expression one expands the two-nucleon wave function into partial waves. Neglecting the coupling of waves of the same orbital angular momentum L but different total angular momentum $\vec{J} = \vec{L} + \vec{S}$ ($\vec{S}$ = total spin) one gets for the various partial wave functions:

$$\psi_{JMLS} = \frac{F_{JLS}(pr)}{pr} Y_{JMLS}(\hat{r}, s_1, s_2) . \qquad (3,65)$$

M is the z-component of $\vec{J}$, s_1 and s_2 denote the spins of nucleon 1 and 2, respectively. Y_{JMLS} are eigenfunctions of the total orbital momentum operator and can be expressed by spherical harmonics, two-particle spin functions and Wigner 3 j-symbols.
The radial wave functions have the following asymptotic behavior:

$$F_{JLS}(pr) \to \sin(pr - \frac{\pi}{2}L + \delta_{JLS}), \quad pr \gg L . \qquad (3,66)$$

As originally proposed by Durand, one can avoid the numerical solution of the Schrödinger equation in the following way: starting from a set of experimentally determined phaseshifts we can adjust certain square well potentials of fixed range to get the right asymptotic behavior (3,66). The depth of the potentials serves as our parameter for fitting the asymptotic solution to the known phaseshifts. As can be seen from Table 2 the choise of the potential range r_o does not strongly influence the magnitude of the final state corrections [20].
As we know the analytical form of the solution for a square well potential, namely a linear combination of a Bessel- and a Neumann function of half integer order we can immediately use (3,66) to find the proper coefficients. To fit the potential depths one uses the continuity condition for the solution and its derivative at the potential boundary. This yields a relation between the phaseshifts and the corresponding potential depth V_{JLS} which can be used to determine the right V_{JLS} for each δ_{JLS}. This quantity enters then the

argument of the sqare well solutions for the inner (potential) region $r \leq r_o$ in the combination $\lambda_{JLS} r$

Table 2. The influence of the final state corrections for different r_o on the quasielastic peak

q/2	p	$\delta I_1/I_1$ (%)		
		$r_o = 1.8$	$r_o = 2.0$	$r_o = 2.2$
1.53	1.53	- 3.39	- 3.93	- 3.73
1.9948	1.9948	- 2.71	- 2.75	- 2.71

q/2	p	$\delta I_2/I_2$ (%)		
		$r_o = 1.8$	$r_o = 2.0$	$r_o = 2.2$
1.53	1.53	- 2.42	- 2.92	- 3.20
1.9948	1.9948	- 2.63	- 2.78	- 2.74

with

$$\lambda_{JLS} = (\vec{p}^{\,2} - 2mV_{JLS})^{\frac{1}{2}} . \qquad (3,67)$$

In analogy to experimental evidence for a hardcore in the n-p-interaction and for consistency with our deuteron wave function with hardcore we choose almost the same hardcore as was used in pot. 9 of Glendenning and Kramer, namely $r_c = 0.4329f$. As the set of phaseshifts we took these from Breit et al. [22]. Inserting the corresponding V_{JLS} into our two-nucleon wave functions our matrix elements contain besides some spherical harmonics and Wigner coefficients the radial integrals K_{JLS}, defined by

$$K_{JLS}(|\vec{p}|, |\vec{q}|) =$$

$$= \frac{1}{|\vec{p}|} \int_{r_c}^{\infty} F_{JLS}(|\vec{p}|r) j_L(\tfrac{1}{2}|\vec{q}|r) u(r) dr \tag{3,68}$$

as far as the non-gradient terms in (3,62) are concerned.
The gradient terms lead to integrals of the form

$$K^k_{JLS}(|\vec{p}|, |\vec{q}|) =$$

$$= \frac{2}{|\vec{q}|} \int_{r_c}^{\infty} \frac{F_{JLS}(|\vec{p}|r)}{|\vec{p}|r} j_k(\tfrac{1}{2}|\vec{q}|r) \frac{\partial}{\partial r}\left(\frac{u}{r}\right) r^2 dr , \tag{3,69}$$

where j_k is a spherical Bessel-function of order k.
Squaring the matrix elements and performing the angular integrations with the help of orthogonality conditions for spherical harmonics and the resulting Wigner coefficients [13] we arrive at the following final expression for I (φ):

$$\begin{aligned} I(\varphi) = \frac{1}{3} \sum_{J,L} & (2J+1) K^2_{JL1} \{(1-\tau)[G_{1p} + (-)^2 G_{1n}]^2 - \\ & - 2\tau [G_{1p} + (-)^L G_{1n}][G_{2p} + (-)^2 G_{2n}]\} + \\ & + \frac{\tau}{3}(2\tan^2\frac{\varphi}{2} + 1) \sum_L \{\tfrac{1}{2}[G_{1p} + (-)^2 G_{1n} + G_{2p} + \\ & + (-)^L G_{2n}]^2 <(3L+4) K^2_{L+1,L,1} + (2L+1) K^2_{LL1} + \\ & + (3L-1) K^2_{L-1,L,1}> + [G_{1p} - (-)^L G_{1n} + \\ & + G_{2p} - (-)^L G_{2n}]^2 (2L+1) K^2_{LLo}\} . \end{aligned} \tag{3,70}$$

For reasons of convenience we have introduced the set of form factors $G_1 = F_1$, $G_2 = F_2/\varkappa$, where $\varkappa$ is the anomalous magnetic moment of the nucleon under consideration.

This expression has to be compared with the Born term $I_o(\varphi)$ without final state corrections which can be found by setting $V_{JLS} = 0$. Before turning to the numerical results we should stress that the whole procedure outlined above strongly depends on the choice of r_c.

This can be seen from Table 3 where we have summarized the relative changes of $I(\varphi)$ at the peak to final state corrections [20]. The corresponding changes in I_1 and I_2 of (3,46) can be fitted by a curve of the form

$$C^{(i)}(q^2) = 1 + 0.01\left(-a_1^{(i)}/q^2 + a_2^{(i)} - a_3^{(i)} q^2 e^{a_4^{(i)} q^2}\right),$$

whose parameters were obtained by a least square fit to the values of Table 3 [19]):

$$a_1^{(1)} = 10.71 \quad a_2^{(1)} = -0.22 \quad a_3^{(1)} = 1.14 \quad a_4^{(1)} = 0.15$$

$$a_1^{(2)} = -7.73 \quad a_2^{(2)} = +1.64 \quad a_3^{(2)} = 0.9 \quad a_4^{(2)} = 0.15 .$$

The same method was used by Hofstadter et al. [23] to represent the result of Nuttall and Whippman [24] who had found somewhat smaller rescattering corrections by solving the uncoupled Schrödinger equation with the Gammel-Thaler potential [25] numerically.

Table 3. Relative changes of $I(\varphi)$ at the quasielastic peak due to final state corrections for various values of r_c

		$\delta I(\varphi)/I(\varphi)$ (%)		
$p = q/2$	r_c	$\varphi = 45°$	$\varphi = 90°$	$\varphi = 135°$
1.3449	0	- 2.34	- 1.75	- 1.12
	.3022	- 3.68	- 3.06	- 2.41
	.4316	- 4.08	- 3.51	- 2.91
	.5610	- 4.42	- 3.94	- 3.44
1.7361	0	- 1.52	- 1.36	- 1.23
	.3022	- 2.65	- 2.49	- 2.36
	.4316	- 3.11	- 3.01	-2.94
	.5610	- 3.59	- 3.59	- 3.58
1.9948	0	- 1.36	- 1.30	- 1.25
	.3022	- 2.26	- 2.21	- 2.17
	.4316	- 2.75	- 2.77	- 2.78
	.5610	- 3.30	- 3.38	- 3.44

Appendix

Connections between quantities in the lab.-system (LS) and in the cm.-system of the outgoing nucleons (NSS):
The final momentum of the nucleons in the NSS may be written (see Fig. 11):

$$p'_c = \begin{pmatrix} E_c \\ |\vec{p}_c| \cos\Theta_c \\ |\vec{p}_c| \sin\Theta_c \end{pmatrix} \qquad n'_c = \begin{pmatrix} E_c \\ -|\vec{p}_c| \cos\Theta_c \\ -|\vec{p}_c| \sin\Theta_c \end{pmatrix} \quad \text{(A1)}$$

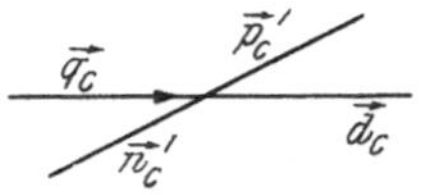

Fig. 11

A Lorentz-transformation along the direction of $\vec{q}_c$ into the L.S. has the form

$$\begin{pmatrix} (1-\beta^2)^{-\frac{1}{2}} & \beta(1-\beta^2)^{-\frac{1}{2}} & 0 \\ \beta(1-\beta^2)^{-\frac{1}{2}} & (1-\beta^2)^{-\frac{1}{2}} & 0 \\ 0 & 0 & 1 \end{pmatrix} \begin{pmatrix} E_c & E_c \\ |\vec{p}_c| \cos\Theta_c & -|\vec{p}_c| \cos\Theta_c \\ |\vec{p}_c| \sin\Theta_c & -|\vec{p}_c| \sin\Theta_c \\ \uparrow & \uparrow \\ p'_c & n'_c \end{pmatrix}$$

$$= (1-\beta^2)^{-\frac{1}{2}} \begin{pmatrix} E_c+\beta |\vec{p}_c| \cos\Theta_c & E_c-\beta|\vec{p}_c| \cos\Theta_c \\ \beta E_c+ |\vec{p}_c| \cos\Theta_c & \beta E_c- |\vec{p}_c| \cos\Theta_c \\ \sqrt{1-\beta^2}\, |\vec{p}_c| \sin\Theta_c & -\sqrt{1-\beta^2}\, |\vec{p}_c| \sin\Theta_c \\ \uparrow & \uparrow \\ p' & n' \end{pmatrix} \quad \text{(A2)}$$

Now we can determine the parameter β, because we have to demand

$$\vec{p}\,' + \vec{n}\,' \stackrel{!}{=} \vec{q} \ , \tag{A3}$$

$$\tfrac{1}{2}\,(\vec{p}\,' - \vec{n}\,') \stackrel{!}{=} \vec{p} \ , \tag{A4}$$

$$p'_0 + n'_0 \stackrel{!}{=} 2m - B + \Delta E \ . \tag{A5}$$

We therefore get

$$\beta = \frac{|\vec{q}\,|}{2}\,(m^2 + \vec{p}_c^{\,2} + \vec{q}^{\,2}/4)^{-\frac{1}{2}} \ . \tag{A6}$$

The LS–scattering angle is given by

$$\cos^2\Theta_c = (m^2 + \vec{p}_c^{\,2})\cos^2\Theta\,(m^2 + \vec{p}_c^{\,2} + \frac{\vec{q}^{\,2}}{4}\,\sin^2\Theta)^{-1} \ . \tag{A7}$$

From these relations and

$$|\vec{p}\,| = |\vec{p}_c|\,\left(\frac{1-\beta^2\sin^2\Theta_c}{1-\beta^2}\right)^{\frac{1}{2}} \tag{A8}$$

the formulae (3,16) and (3,17) follow directly.

Literature

[1] R. Hofstadter, F. Bumiller, M.R. Yearian, Rev. Mod. Phys. 30, 482 (1958).

[2] S.D. Drell, F. Zachariasen, Electromagnetic Structure of Nucleons, Oxford:University Press (1961).

[3] M. Gourdin, Nuovo Cim. 28, 533 (1963).

[4] F. Gross, Phys. Rev. 134, B 405 (1964), 140, 410 (1964).

[5] F. Gross, Phys. Rev. 142, 1025 (1966).

[6] R.M. Renard, J. Tran Than Van, M. Le Bellac, Nuovo Cim. 38, 565 (1965).

[7] J. McGee, Thesis, Yale University (1965).

[8] D. Braess, G. Kramer, Proc. of the Int. Symposium on Electron and Photon Interactions at High Energies, Hamburg (1965), Vol. II , p. 60.

[9] L. Durand, Phys. Rev. 123, 1393 (1961).

[10] K. Hölzl, G. Saller, P. Urban, Acta Phys. Austr. 19, 168 (1964).

[11] V.Z. Jancus, Phys. Rev. 102, 1586 (1956).

[12] L. Durand, Phys. Rev. 115, 1020 (1959).

[13] K. Hölzl, Thesis, Univ. Graz (1963).

[14] L.L. Foldy, S.A. Wouthuysen, Phys. Rev. 78, 29 (1950).

[15] K.M. McVoy, L. van Hove, Phys. Rev. 125, 1034 (1962).

[16] K. Hölzl, G. Saller, P. Urban, Phys. Lett. 10, 120 (1964).

[17] P. Breitenlohner, Proc. of the Int. Symposium on Electron and Photon Interactions at High Energies, Hamburg (1965), Vol. II, p. 78.

[18] P. Breitenlohner, Acta Phys. Austr. 22, 217 (1966).

[19] P. Kocevar, Z. Phys. 209, 457 (1968).

[20] P. Breitenlohner, K. Hölzl, P. Kocevar, Proc. of the Int. Symposium on Electron and Photon Interactions at High Energies, Hamburg (1965), Vol. II, p. 73.

[21] N.K. Glendenning, G. Kramer, Phys. Rev. Lett. 7, 471 (1961).

[22] M.H. Hull, K.E. Lassila, H.M. Ruppel, F.A. McDonald, G. Breit, Phys. Rev. 122, 1606 (1961).

[23] E.B. Hughes, T.A. Griffy, M.R. Yearian, R. Hofstadter, Phys. Rev. 139, B458 (1965).

[24] J. Nuttall, M.L. Whippman, Phys. Rev. 130, 2495 (1963).

[25] J.G. Gammel, R.M. Thaler, Phys. Rev. 107, 291, 1337 (1957).

IV. Calculation of Nucleon Form Factors in Dispersion Theory

The dispersion-theoretic treatment is based on certain statements about the analytic properties of the form factors considered functions of a complex variable z.

Fig. 12 shows the relevant diagram we have to deal with

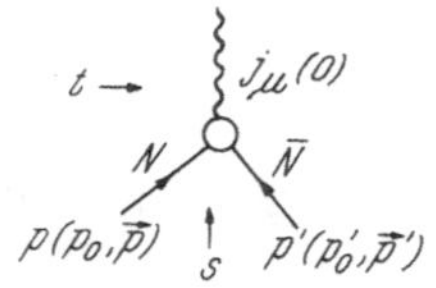

Fig. 12

The hatched bubble contains contributions of various particles consistent with the conservation laws known. These contributions provide for a non-local interaction of $j_\mu(o)$ with the nucleon; in other words, the nucleon must not be considered a point particle. This fact is taken into account by the so-called form factors.

To get some feeling of the dispersion theoretic treatment of form factors we first ignore complications due to spin and isospin. We define three functions [1]:

$$(4_{p_0 p_0'})^{\frac{1}{2}} \langle p' | j(o) | p \rangle = I(t) \quad t = (p'-p)^2 \tag{4,1a}$$

$$(4_{p_0 p_0'})^{\frac{1}{2}} \langle p p', \text{ out } | j(o) | o \rangle = J(s) \quad s = (p+p')^2 \tag{4,1b}$$

$$(4_{p_0 p_0'})^{\frac{1}{2}} \langle p p', \text{ in } | j(o) | o \rangle = K(s) \quad s = (p+p')^2 \tag{4,1c}$$

(4,1a) is the usual definition of a physical form factor. We shall show that all these functions are definite boundary values of a common function F (z). (Rez = s, t).

To obtain some relations between I (t), J (s) and K (s) we next use space reflection P and time reflection T assuming that j (o) is invariant under P and T (which is the case in strong and electromagnetic interactions) :

$$J(s) = (4_{p_0 p'_0})^{\frac{1}{2}} \langle pp', \text{ out } | (TP)^{-1} (TP) j(o) (TP)^{-1} (TP) | o \rangle =$$

$$= (4_{p_0 p'_0})^{\frac{1}{2}} \langle pp', \text{ in } j^c(o) | o \rangle =$$

$$= (4_{p_0 p'_0})^{\frac{1}{2}} \langle pp', \text{ in } | j(o) | o \rangle^*$$

C means complex conjugation since TP is antiunitary .

$$J(s) = K^*(s) \tag{4,2}$$

Similarly we get

$$I(t) = I^*(t) \tag{4,3}$$

Of course (4,2) and (4,3) are only valid in the physical region

$$s > \text{ some threshold energy}$$
$$t < 0 \tag{4,4}$$

Furthermore using the retarded commutator contraction scheme (see also the derivation of equation (4,11)) and contracting the particle p we find the well known crossing relation

$$I(p, p') = K(-p, p') \tag{4,5}$$

It shows that

$$I(t) = K(s) \tag{4,6}$$

Equation (4,6) can only be given meaning by a definite analytic continuation (see equation (4,4)). Notice that the Mandelstam variables s, t, u satisfy the relation

$$s + t + u = \Sigma m_i^2$$

Thus the reader should always keep in mind that for u fixed the variable t may be replaced by s through the relation above and vice versa.

We now establish a dispersion representation for K(s) to verify this continuation

$$K(s) = (4_{p_0 p'_0})^{\frac{1}{2}} \langle p p', \text{ in } | j(o) | o \rangle =$$

$$= (4_{p_0 p'_0})^{\frac{1}{2}} \langle p' | a_p(\text{in}) j(o) | o \rangle =$$

$$= i (2_{p'_0})^{\frac{1}{2}} \int_{x_0 \to -\infty} d^3x \, e^{ipx} \overleftrightarrow{\partial}_o \langle p' | \varphi(x) j(o) | o \rangle \tag{4,7}$$

In equation (4,7) we contracted the particle p and used

$$a_p(\text{in}) = i (2p_0)^{-\frac{1}{2}} \int_{x_0 \to -\infty} d^3x \, e^{ipx} \overleftrightarrow{\partial}_0 \varphi(x) \ . \tag{4,8}$$

We replace $\varphi(x) j(o)$ by

$$\Theta(-x) [\varphi(x), j(o)] = - \Theta(-x) [j(o), \varphi(x)]$$

and employ the identity

$$\int_{x_0 \to -\infty} d^3x = \int d^4x \, \partial_0 + \int_{x_0 \to +\infty} d^3x \ . \tag{4,9}$$

Now we get

$$K(s) = i (2p'_0)^{\frac{1}{2}} \int d^4x \, \partial_0 \{ e^{ipx} \overleftrightarrow{\partial}_o \langle p' | \Theta(-x) [j(o), \varphi(x)] | o \rangle \} =$$

$$= i (2p'_0)^{\frac{1}{2}} \int d^4x \, \{ e^{ipx} [\partial_0^2 \langle | \ | \rangle] - [\partial_0^2 e^{ipx}] \langle | \ | \rangle \} \ . \tag{4,10}$$

e^{ipx} is a solution of the Klein-Gordon equation. Thus we may replace $\partial_0^2 e^{ipx}$ by $(\nabla^2 - m^2) e^{ipx}$. By means of two partial integrations we transfer ∇^2 to act on the matrix element (the thereby occurring surface terms vanish).

Finally we obtain

$$K(s) = i\,(2p_0')^{\frac{1}{2}} \int d^4x\, e^{ipx}\, K_x <p'\,|\Theta(-x)\,[\,j\,(o),\, \varphi(x)\,]\,|\,o>$$

or

$$K(s) = i(2p_0')^{\frac{1}{2}} \int d^4x\, e^{ipx} <p'\,|\,\Theta(-x)\,[\,j\,(o),\; \eta(x)\,]\,|\,o> + K'(s) \tag{4,11}$$

K_x ... Klein - Gordon operator

η ... source operator defined by $K_x \varphi(x) = \eta\,(x)$

K' (s) includes the terms which arise from differentiating the Θ-function. One can show that

$$K'(s) = \Sigma B_n\, s^n \tag{4,12}$$

where B_n are constants (see e.g. [1])

This term does not influence the analytic behavior of K (s), therefore we drop it.

Now we consider K (s) in a particular Lorentz-frame

$$\vec{p}\,' = o, \quad p_0' = m, \quad |p'> = |\,m>\,; \quad p_0 \equiv \omega \tag{4,13}$$

$$s = 2\,(m^2 + m\,\omega)$$

As a function of ω K (s) may be written as

$$K(s) = i\,(2\,m)^{\frac{1}{2}} \int d^4x\, e^{i\,\omega x_0 - i\vec{p}\vec{x}}\,.$$

$$\cdot\, \Theta(-x_0) <m\;|\;[\;j\,(o)\,\eta\,(x)\;]\;|\,o> \tag{4,14}$$

We expect that the integral converges for Im $\omega < o$.

This is a conjecture rather than an established result because $e^{-i\vec{p}\vec{x}}$ may well diverge for certain ω (a rigorous proof is for instance given by [2], [3]).

We see that $K(\omega)$ is analytic in the lower complex ω-plane excluding the real axis. Applying the Cauchy-theorem we get ($\omega =$ = Rez)

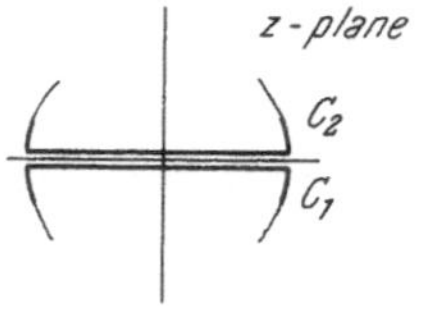

Fig. 13

$$K(z) = (2\pi i)^{-1} \int_{C_1} \frac{dz' K(z')}{z' - z} . \tag{4,15}$$

If $K(z') \to o$ for $z' \to \infty$ we may neglect the semicircle:

$$K(z) = - (2\pi i)^{-1} \int_{-\infty}^{+\infty} \frac{d\omega' K(\omega')}{\omega' - z} . \tag{4,16}$$

A similar consideration defines J (z) to be analytic in the upper half plane

$$J(z) = (2\pi i)^{-1} \int_{C_2} \frac{dz' J(z')}{z' - z} \tag{4,17}$$

or

$$J(z) = (2\pi i)^{-1} \int_{-\infty}^{+\infty} \frac{d\omega' J(\omega')}{\omega' - z} . \tag{4,18}$$

Adding (4,16) and (4,18) we get a new function

$$F(z) \equiv J(z) + K(z) = (2\pi i)^{-1} \int_{-\infty}^{+\infty} d\omega' \frac{J(\omega') - K(\omega')}{\omega' - z} . \tag{4,19}$$

We cannot yet deduce the analytic properties of F (z) because J (z) and K (z) are not well defined for real z.
However, we may apply equations (4,2), (4,3) and (4,6) and obtain

$$J(s') - K(s') = K^*(s') - K(s') = I^*(s') - I(s') = o \quad \text{for } s' < o$$

or

$$J(\omega') - K(\omega') = o \qquad \text{for } \omega' < -m$$

and

$$F(z) = (2\pi i)^{-1} \int_{-m}^{\infty} \frac{J(\omega') - K(\omega')}{\omega' - z} \, d\omega' \; . \tag{4,20}$$

F (z) now represents a function analytic in the entire complex z-plane except for a cut from − m to infinity.
The reality of $F(\omega')$ in the region $\omega' < -m$ is not surprising since there the physical form factor is actually defined (see p. 81).
It is possible to push the branch point from $\omega' = -m$ to $\omega' = m$.
For this purpose we substitute the representation (4,14) for the integrand, insert a complete set of intermediate states and perform the integrations. We arrive at $(p = (\omega', \vec{p}))$

$$[J(\omega') - K(\omega')] =$$

$$= -i(2\pi)^4 (2m)^{\frac{1}{2}} \sum_n \{ \langle m | j(o) | n \rangle \langle n | \eta(o) | o \rangle \delta(p+n) -$$

$$- \langle m | \eta(o) | n \rangle \langle n | j(o) | o \rangle \delta(p+p'-n) \} \; . \tag{4,21}$$

In the first sum only the single particle states contribute. It vanishes because (e.g. [1])

$$\langle n | \eta(o) | o \rangle = 0 \; .$$

An inspection of the second sum shows that the first permissible intermediate state is a two-particle state. The δ-function

$\delta(\omega' + m - n_o) = \delta(\omega' + m - 2m)$ gives

$$\omega' = m \quad . \tag{4,22}$$

We change the variable from ω' to s' (we leave the special Lorentz frame above) and finally obtain

$$F(z) = \int_{4m^2}^{\infty} \frac{\sigma(s')}{s'-z}\, ds' \tag{4,23}$$

where

$$\sigma(s') =$$

$$= (2\pi)^3 (2p_o')^{\frac{1}{2}} \sum_n \langle p' \mid \eta(o) \mid n \rangle \langle n \mid j(o) \mid o \rangle\, \delta(p+p'-n) \tag{4,24}$$

$$s = (p + p')^2$$

We see that $F(z)$ is analytic in the entire complex s-plane with cuts along the positive real axis, the first beginning at $s = 4m^2$.
It is now easy to relate our functions $I(t)$, $J(s)$, $K(s)$ to $F(z)$:

$$J_\epsilon(s) = \lim_{z \to s + i\epsilon} F(z)$$

$$K_\epsilon(s) = \lim_{z \to s - i\epsilon} F(z) \tag{4,25}$$

$I(s) = F(s)$ for $s < 4m^2$ in both limits

Fig. 14

Comparing (4,20) with (4,23) and (4,24) we see that

$$\sigma(s) = \frac{2i}{\pi} [J(s) - K(s)] =$$

$$= \frac{2i}{\pi} [F(s+i\epsilon) - F(s-i\epsilon)] = \frac{1}{\pi} \text{Im } F(s) \quad . \qquad (4,26)$$

If we now include the spin the most general ansatz for the photon-nucleon vertex is (see (2,6))

$$\Gamma_\mu = (4 p_0 p'_0)^{\frac{1}{2}} <p' | J_\mu(o) | p> =$$

$$= e\bar{u}(p') \{ F_1(t) \gamma_\mu + i \frac{(p'-p)^\nu}{2m} \sigma_{\mu\nu} F_2(t) \} u(p)$$

$$t = (p'-p)^2 \qquad (4,27)$$

Now our task is to express F_1 and F_2 in forms not containing spinors. Each component of Γ_μ satisfies a dispersion relation (4,13). We construct the scalar products

$$A' \equiv \bar{u}(p') \gamma^\mu u(p) \Gamma_\mu$$

$$B' \equiv -\bar{u}(p') i (p'-p)^\lambda \sigma^\mu_\lambda u(p) \Gamma_\mu \qquad (4,28)$$

The spinor combinations in (4,28) are analytic in the cut t-plane. Thus A' and B' have the same analytic behavior as Γ_μ. We sum over the spins and evaluate the traces. Then the form factors F_1 and F_2 are

$$e F_1 = m^2 \{ -A [\frac{t}{2m^2} - 4] + 6B \} / 2(4m^2 - t)^2$$

$$e F_2 = -m^2 \{ -6tA + 4(2m^2 + t) B \} / 2t(4m^2 - t)^2 \qquad (4,29)$$

where

$$A = \sum_{\text{spins}} A' \qquad \text{and} \qquad B = \sum_{\text{spins}} B'$$

Obviously F_1 and F_2 have the same analytic properties as A and B.

Furthermore, F_1 (F_2) may be singular at $t = 4m^2$ ($t = 0, 4m^2$). These singularities are known as kinematic singularities.

We know that A and B have a branch point, the normal threshold at $t = 4m^2$. This threshold is a boundary point of the physical region. Here F_1 and F_2 represent an observable matrix element which cannot become infinite in the physical region. So A and B must have a zero strong enough to keep F_1 and F_2 finite here. Furthermore, F_2 (o) cannot be singular because its value is here the measurable Pauli anomalous moment of the particle. We conclude that F_1 and F_2 are free of kinematic singularities and may be represented by a dispersion relation. Our conclusions are valid both for the proton form factors $F^p_{1,2}$ and the neutron form factors $F^n_{1,2}$. It is convenient to introduce the isospin formalism and to express $F^p_{1,2}$ and $F^n_{1,2}$ by isoscalar and isovector form factors $F^s_{1,2}$ and $F^v_{1,2}$

$$F^p_{1,2} = F^s_{1,2} + F^v_{1,2} \tag{4,30}$$

$$F^n_{1,2} = F^s_{1,2} - F^v_{1,2}$$

In our case the lowest intermediate states are (2π) and (3π) states

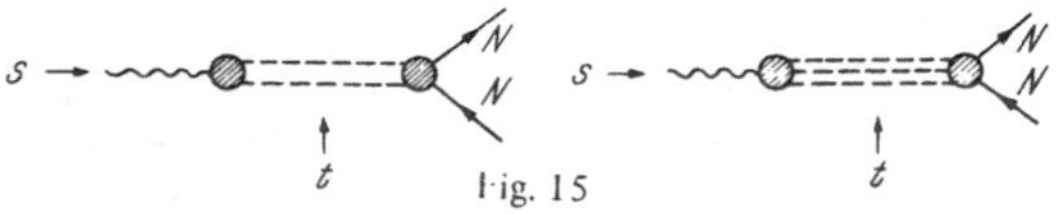

Fig. 15

We take only these states for two reasons:

a) These states can be calculated

b) If $\sigma(s')$ has the peak at some value s_p the contribution as $s = s_p$ will dominate and the form factors will have the form [4]

$$F(s) \simeq \frac{s_p}{s_p - s} \quad .$$

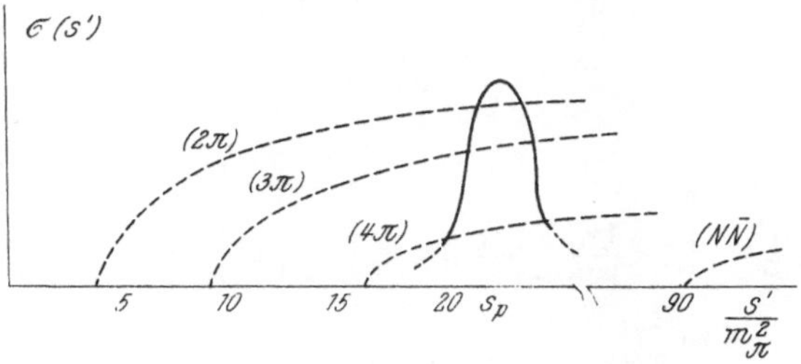

Fig. 16

To explain the experimental values for F^S and F^V we need a peak at

$$s_p \simeq 20\, m_\pi^2 \quad .$$

We see that states with higher energies do not contribute in this region. We may even simplify the situation by taking conservation laws into account. The photon has the quantum numbers $C = -1$, $J^P = 1^-$. Thus the G-partity for the meson states is

$$G = C(-1)^I = (-1)^{I+1} = (-1)^{n_\pi}$$

I . . . . isospin

For F^V we have $I = 1$ and n_π even.
For F^S we get $I = 0$ and n_π odd.
We obtain from (4,23) and (4,26)

$$F^V(s) \equiv \lim_{\epsilon \to 0} F^V(s + i\epsilon) = \lim_{\epsilon \to 0} \frac{1}{\pi} \int_{4m_\pi^2}^{\infty} \frac{\operatorname{Im} F(s')}{s' - s - i\epsilon}\, ds'$$

$$F^S(s) \equiv \lim_{\epsilon \to 0} F^S(s+i\epsilon) = \lim_{\epsilon \to 0} \frac{1}{\pi} \int_{9m_\pi^2}^{\infty} \frac{\operatorname{Im} F(s')}{s'-s-i\epsilon} ds'$$

or simply

$$F^V(s) = \frac{1}{\pi} \int_{4m_\pi^2}^{\infty} \frac{\operatorname{Im} F s')}{s'-s-i\epsilon} ds' \tag{4,31}$$

$$F^S(s) = \frac{1}{\pi} \int_{-9m_\pi^2}^{\infty} \frac{\operatorname{Im}(s')}{s'-s-i\epsilon} ds' \tag{4,32}$$

We neglect the (4π) states and suppose that the (2π) and (3π) states represent the whole spectrum. In the following we consider the isovector form factor (the scalar form factor may be treated in the same manner). We assume that the contribution to the imaginary part of F (s) is given by Fig. 17.

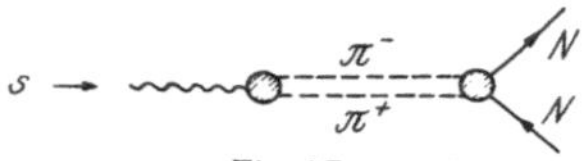

Fig. 17

There are two factors:

a) the coupling of a time like photon to a $\pi^+\pi^-$ - pair,

b) the amplitude of the $N\bar{N}$ annihilation into two π's.

This amplitude lies in the unphysical region. Thus we need an analytic continuation from $s' = 4m^2$ to the region $4m_\pi^2 \leq s' \leq 4m^2$;
We find the following behavior (assuming a point interaction between γ and (2π)).

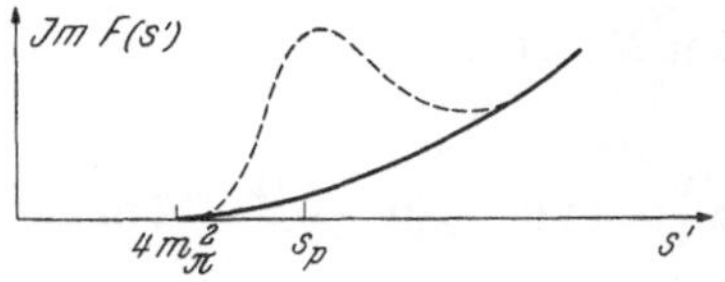

Fig. 18

To explain the experimental facts a modification (dotted line) is necessary. A simple dimensional exercise yields the isovector radius at the threshold $\langle r^2 \rangle_{2\pi} = \frac{1}{4 m_\pi^2} = 0,25\, m_\pi^{-2}$. This value is in remarkable conformity with the observed radius $\langle r^2 \rangle_{exp} = 0,2\, m_\pi^{-2}$. Thus we see that qualitatively we are on the right track. But quantitatively we have some troubles. One can hardly believe that the spectral function reaches its maximum near threshold (phase space limitations !). Since the amplitude b) varies only slowly with s something drastic must happen to the (2π) system to resolve the contradiction inherent in the conformity with the experimental value and the fact that the spectral function attains its maximum somewhere in the region $s' > 4 m_\pi^2$. The simplest possibility is to suppose a strong attractive interaction between the two pions. The resulting resonance must be a vector particle since the photon carries the total momentum J = 1. The prediction of this resonance - the well known ρ-meson ($J^{PG} = 1^{-+}$, m = 760 MeV, Γ = 128 MeV) - was one of the most brilliant successes of dispersion theory.

The experimental results for the nucleon form factors have been analyzed in terms of the known vector mesons. In order to keep the discussion transparent we begin by considering only a single resonance in each isovector form factor, taking the ρ-meson for definiteness. As we have seen the ρ-resonance causes a peak in the (2π) contribution to the absorptive part of F_i^V.

In case of electron scattering experiment the F_i^V are required for negative s. This region is far away from the location of the resonance. So we neglect the width of the resonance and replace it by a δ-function. Inserting the ρ intermediate state into the expression for Im F_i we get

$$e\bar{u}(p')\{\gamma_\mu \operatorname{Im} F_1^V + (i\sigma_{\mu\nu}(p'-p)^\nu/2m)\operatorname{Im} F_2^V\}u(p) =$$

$$= \pi(2p_0)^{\frac{1}{2}}\sum_\rho <p'|\eta|\rho><\rho|j_\mu|o>\delta(p+p'-\rho)u(p) \quad . \tag{4,33}$$

Note that in contrast to equation (4,24) we are now dealing with spin particles. The first factor introduces the two ρNN - coupling constants f_1^V and f_2^V defined by

$$(2p_0)^{\frac{1}{2}}<p'|\eta|\rho>u(p) =$$

$$= \bar{u}(p')\{\gamma_\lambda f_1^V + i(p_1-p_2)^\alpha \sigma_{\lambda\alpha} f_2^V\}(2\omega_\rho)^{-\frac{1}{2}}\epsilon_\lambda(\rho)u(p)$$

$$(p'-\rho)^2 = m^2 \tag{4,34}$$

The second factor may be written as

$$<\rho|j_\mu|o> = (2\omega_\rho)^{-\frac{1}{2}}\epsilon_\mu(\rho)m_\rho^2 f_{\rho\gamma} \tag{4,35}$$

$\epsilon_\gamma(\rho)$ ... polarization vector of ρ

$f_{\rho\gamma}$ .... a dimensionless coupling constant .

In deriving (4,35) we assumed that $\rho^\lambda j_\lambda = o$.
We insert (4,34) and (4,35) into (4,33) and get

$$e \operatorname{Im} F_1^V = \pi f_1^V f_{\rho\gamma} m_\rho^2 \delta(s-m_\rho^2) \tag{4,36a}$$

$$e \operatorname{Im} F_2^V = 2m\pi f_2^V f_{\rho\gamma} m_\rho^2 \delta(s-m_\rho^2) \tag{4,36b}$$

We identify (4,36b) with $\operatorname{Im} F^V$ in (4,32) and use for (4,36a) a once substracted dispersion relation to fit the constant, which now appears with the fundamental constant e. The resulting expressions for F_1^V and F_2^V are

$$e F_1^V(z) = \left[\frac{e}{2} - f_1^V f_{\rho\gamma}\right] + \frac{f_1^V f_{\rho\gamma} m_\rho^2}{m_\rho^2 - z} \tag{4,37a}$$

$$e F_2^V(z) = \frac{2 m m_\rho^2}{m_\rho^2 - z} f_2^V f_{\rho\gamma} \; . \tag{4,37b}$$

These equations are sometimes called Clementel-Villi formula. Evidently we get

$$e F_1^{V'}(o) = \frac{f_1^V f_{\rho\gamma}}{m_\rho^2} \; ; \quad \frac{F_2^{V'}(o)}{F_2^V(o)} = \frac{1}{m_\rho^2} \; . \tag{4,38}$$

If m_ρ^2 is taken from experiment then $f_1^V f_{\rho\gamma}$ determines F_1^V and vice versa; F_2^V is uniquely fixed in terms of m_ρ^2 and $F_2^V(o)$

$$F_2^V(o) = 1{,}85 \; .$$

But F_2^V and f_2^V themselves cannot be calculated without going beyond this one pole approximation.

Frazer and Fulco (see e.g. [5]) fitted the resonance position, i.e. m_ρ^2, to reproduce the observed form factors. They had already fixed $f_{\rho\gamma}$ implicitly by their parametrization of the pion form factor (By means of the strong interactions between the two pions they have a definite structure which introduces a form factor for the vertex 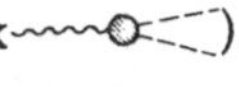).

Thus Frazer and Fulco actually determined f_1^V and f_2^V. Considering the isoscalar form factors there are now two low lying resonances, the ω-meson ($J^{PG} = 1^{--}$, $m = 780$ MeV, $\Gamma = 10$ MeV) and the φ-meson ($J^{PG} = 1^{--}$, $m = 1020$ MeV, $\Gamma = 3$ MeV) which have to be taken into account. Their contributions to $F_i^S(z)$ in equation (4,37) simply constitute additions. In order to compare the form factors with the experiment it is convenient to use the electric and the magnetic form factors G_E and G_M defined in equations (2,11). Since they are linear combinations of F_1 and F_2 it is evident that $G^s_{E,M}$ and $G^v_{E,M}$ exhibit the same functional form as F in (4,37). Approximating the intermediate states again by the three meson resonances (3-pole fit) we get equation (2,37) which can be fitted to the experimental data. Notice that in equation (2,37) m^2_ρ is taken as a parameter rather than a fixed mass. This is necessary to compensate the error which arises by considering the ρ-meson as a stable particle; actually it has a large width.

Literature

[1] G. Barton, Dispersion Techniques in Field Theory, New York: W.A. Benjamin (1965).

[2] M.L. Goldberger, Dispersion Theory and Elementary Particles, New York: John Wiley & Sons (1961).

[3] M.L. Goldberger, K.M. Watson, Collision Theory, New York: John Wiley & Sons (1964).

[4] R. Rodenberg, unpublished lectures 1964.

[5] W.R. Frazer, Dispersion Ralations, ed. Screaton. Edinburgh: G.R. Oliver & Boyd (1961).

Part II

RADIATIVE CORRECTIONS

Introduction

In connection with an introduction to the subject - radiative corrections to high-energy electron scattering - we shall give here a summary of the standard literature so that in the following chapters further references will be omitted.

First an explanation of the term "radiative corrections": The interaction between electrons and photons is appropriately described by quantum electrodynamics; in this framework, however, it is necessary to employ the perturbation theory where higher approximations correspond physically to the insertion of further radiation quanta into the basic process under consideration. Mathematically, these higher approximations show up as additional contributions to the original transition probability; these are the so-called radiative corrections. Also from the classical point of view it is intelligible that any scattering process involving charged particles is accompanied by radiation quanta since an electric current is always surrounded by an electromagnetic field. It is therefore evident that the above-mentioned corrections are not a pure quantum effect but are an already classically occurring phenomenon.

Another important concept is the "infrared divergence"; it is closely connected with the radiative corrections although these divergences occur only in quantum electrodynamical calculations and have no classical analogon. In quantumelectrodynamics we experience the following difficulties: two different types of divergences have their origin in the integration over momenta of the additional photons.

The first type is due to the upper limit; these ultraviolet divergences can be eliminated by means of the renormalization theory. The divergences from the lower limit, however, remain (e.g. from integrals of the type $\int_{0} d\omega/\omega$). These infrared divergences are not only connected with the exchange of virtual photons but also occur in the emission of real bremsstrahlungs quanta. In the description of realistic experiments we have to take into account also the latter possibility since due to the inherent uncertainties in observation a given process cannot be distinguished from the inelastic ones with real photons of sufficiently low energy. It is very interesting that the infrared divergences disappear (they cancel out exactly) if we combine elastic and inelastic contributions. This fact is a hint that the decomposition into real and virtual photons is not permitted within the limit of vanishingly small energy. The infrared divergences therefore appear artificially due to the application of the perturbation theory to photons of extremely long wavelength; they do not occur at all in the classical description of these quanta.

This conclusion was obtained by Bloch and Nordsieck [1] in their famous work on the radiation field of the electron where for the first time the infrared divergences were avoided. This was accomplished by a classical approximation (substitution of c-numbers for Dirac matrices, neglect of antiparticles) for the interaction with the field of lang-wavelength quanta. The most important results of this work were the conclusions that the radiated intensity per frequency interval remains finite in the limit $\omega \to o$, and that the average number of emitted quanta goes to infinity and that therefore these infinite number of quanta gives rise to a finite correction factor (which Bloch and Nordsieck calculated to be one) whereas the probability for the emission of a finite number of soft ($\omega \to 0$) photons vanishes.

This work provided for the key to the solution of the infrared problem but had also some drawbacks. It had for instance been overlooked that in the treatment the photons could carry with them an arbitrarily large amount of energy which is in contradiction with energy conservation. Pauli and Fierz [2] included then energy conservation in their calculation but arrived at the obviously incorrect result that the correction factor and therefore also the corrected transition probability vanishes. This result came from a newly appearing ultraviolet divergence due to the integration over the photon spectrum. It took a long time until the problem was solved in a satisfactory way. Roughly speaking, difficulty had its origin in the fact that on the one hand one used an approximation scheme for the soft photons, valid for small frequencies only, whereas on the other hand the integration was extended to $\omega \to \infty$ where the assumption for soft photons was not fulfilled any more. Therefore one was forced to introduce a cut-off which guaranteed the validity of the approximation.

Jauch and Rohrlich [3] solved the problem in a satisfactory manner. Generally speaking, this work may be regarded as the basis for all further results concerning the avoidance and the removal of infrared divergences, respectively, and also serves as a survey of the history of the problem worth reading. The result of a simple model calculation gives a correction factor $b = 1 + \delta$ where δ is of the order of magnitude of the finestructure constant α multiplied by a factor containing the experimental uncertainties in measurement and the energies of the particles. For high energies and accurate measurements the radiative correction δ can reach considerable values of the order of 30 %. The infrared divergences disappeared, there remained, however, significant corrections which still are often called "infrared" contributions.

Starting from Jauch and Rohrlich essentially two different methods have been developed: one group makes use of quantum-electrodynamics and the cancellation of the infrared divergences. A pioneering work in this respect is the one by Schwinger [4], where an explicit first order calculation is carried out. In the work of Jauch and Rohrlich this result has been generalized and it could be proved that the divergences cancel in any order of the perturbation series. The methods have constantly been refined and it was finally possible to sum completely over all infrared contributions (of soft photons); the most important paper in this connection is the one by Yennie, Frautschi and Suura [5]. The difficulties arising in the application of this method are mainly mathematical in nature, as e.g. the determination of the phase space of soft photons from the experimental conditions and the corresponding integrations which sometimes have to be carried out numerically. Also the final results are relatively lengthy and not easily intelligible.

The second group takes as its starting point the fact that the soft quanta can be described by means of a classical approximation avoiding the occurrence of divergences. Also here the summation over all soft photons can be executed ([6], [7]) and leads to a result similar to the one achieved by the other method; the two final formulas cannot be compared analytically in every detail, they lead, however, to practically identical numerical values. Touschek [8] then proceeds from the fact that the photons are governed classically by a Poisson distribution, and develops a surprisingly simple technique for the evaluation of the infrared terms. This method, however, seems to work well only under special experimental conditions, sharp resolution and resulting large δ. The contributions of hard photons, which are of so high an energy that the classical approximation breaks down, have to be calculated separately and again by means of quantumelectrodynamics. Certain difficulties also arise in connection with fixing the limiting energy between soft and hard photons.

Now a comment on the significance of radiative corrections for the experiments: it is obvious even from the above quoted (extremely large) numerical value of the correction. Increasing experimental accuracy and the use of constantly higher energies stress the need for a more exact knowledge of radiative corrections. A serious hindrance towards this goal is that the corrections strongly depend on the special experimental situation, therefore different calculations and approximations have to be employed for different experiments. These problems have been discussed extensively by Meister and Yennie [9] and Tsai [10]. There one may find other details which we do not discuss here. In these works a "recipe" for a large class of electron scattering experiments is given of how to calculate radiative corrections in a relatively simple way. The importance of radiative corrections may be deduced also from the number of talks on this subject presented at the International Symposium on Electron and Photon Interactions at High Energies in Hamburg (1965); we especially want to mention the contributions made by Tsai (p. 387), Lomon (p. 94), Kohaupt (p. 91) and Bartl (p. 86).

Finally we want to give a brief survey of the problems dealt with in the following chapters. We start with the intuitive classical treatment, then proceed to the quantumelectrodynamical method and finally discuss their application on some examples which we take from our own works [11].

It is obvious that thereby we could not cover all the problems connected with radiative corrections; for instance we totally omit the corrections to weak interactions. Neither can we discuss here the problems involving strong interactions. The subject has such a vast range that in our examples we have to restrict ourselves to a class of very special electron scattering experiments. However, we have tried to present in detail in order to enable further study of the special problems in the literature. For the sake of completeness we finally list further references which attack the subject from different points of view ([13]) or are of more recent date ([14], [15]). With this literature in hand a more detailed study of the problems discussed here can be accomplished. Similar review articles may be found in the proceedings of various summer schools, as e.g. an article by Yennie in the Brandeis-Lectures 1963.

I. Classical Radiation of Long-Wavelength Photons

1. Introduction

In the description of scattering processes involving charged particles electric currents are conveniently introduced and since such currents are surrounded by electromagnetic fields it is necessary to include also the effect of a certain number of photons accompanying the process. This purely classical phenomenon may of course be treated by classical means with the advantage of avoiding the difficulties inherent in the perturbation expansion of QED for low-energy photons. In the classical treatment no distinction is made between real and virtual photons; exactly this unphysical splitting causes the quantum-electrodynamical description to fail for long-wavelength photons.

Certain conditions, however, must be met in order to apply the classical description. The reaction of the photons on the (particle-) current must be negligible and, furthermore, the wavelength of the photons has to be so long as not to measure the details of the scattering process. This condition, for the standard case of the scattering of an electron in an external field, is expressed mathematically as

$$\hbar\omega \ll E_{kin} \;;\; \frac{\hbar\omega}{c} \ll \Delta p \;; \tag{1,1}$$

where E_{kin} and Δp are the electron's kinetic energy and change in momentum respectively. In addition

$$r_0 \ll \lambdabar \;\left(\text{or } \frac{e^2\omega}{mc^3} \ll 1\right) , \tag{1,2}$$

that is, the classical electron radius should be small compared to the photon wavelength.

If, for reasons of simplicity, we assume the electron to be non-relativistic $(v \ll c)$ a short classical calculation yields the number of accompanying long-wavelength photons: the average number is given by the ratio of the intensity of radiation per unit frequency interval and the energy of the quanta as

$$\bar{n} = \frac{1}{\hbar\omega}\frac{dI_\omega}{d\omega} , \tag{1,3}$$

in the limit $\omega \to 0$. The intensity of dipole radiation

$$dI_\omega \sim |\ddot{\vec{\delta}}_\omega|^2 d\omega ,$$

is proportional to the square of the Fourier-transform of the second time-derivative of the dipole moment

$$\ddot{\vec{\delta}}_\omega = \frac{1}{(2\pi)^{\frac{1}{2}}} \int_{-\infty}^{+\infty} \ddot{\vec{\delta}} \, e^{i\omega t} \, dt .$$

With the dipole moments δ_1, δ_2, before and after the scattering process respectively, we get for small ω

$$\lim_{\omega \to 0} (\ddot{\vec{\delta}}_\omega) \simeq \frac{1}{(2\pi)^{\frac{1}{2}}} (\dot{\vec{\delta}}_2 - \dot{\vec{\delta}}_1) .$$

Since these moments can be expressed by means of the relevant electron densities as

$$\vec{\delta} = \int \rho \vec{r} \, d\tau ,$$

$$\dot{\vec{\delta}} = \frac{\partial}{\partial t} \int \rho \vec{r} \, d\tau = e \vec{v} ,$$

we see that the intensity of radiation for small ω

$$\lim_{\omega \to 0} dI_{\omega} \sim e^2 (\vec{v}_2 - \vec{v}_1)^2 \, d\omega ,$$

is finite and therefore the average number of "soft" photons approaches infinity.

Another conclusion from this result is that the probability for the emission of a single photon (or a finite number of photons) must vanish. Since the scattering process certainly is observable its associated transition probability must be finite, and so is, as may be seen from the foregoing calculation, the probability of having on the average an infinite number of soft photons accompanying the process. The probability for the emission of one soft photon is proportional to $\ln (E_{kin} / \omega)$ (see bremsstrahlung spectrum) and for the independent emission of n photons $(\ln E_{kin}/\omega)^n$. The ratio of the probability for many photons to the probability for one photon goes to infinity as ω goes to zero, therefore the probability for one photon (or a finite number of photons) must vanish since the probability for an infinite number of photons was found to be finite.

Under the conditions (1,1) and (1,2) the calculation of the probability that an (electron-) current emits a number (n) of soft photons can be done classically. Since the different emission processes are assumed to be statistically independent the probability for the emission of soft photons can be represented by a Poisson-distribution. The next section deals with the emission problem in more detail; in the following paragraphs we shall give a simple derivation of the Poisson distribution together with some general consequences.

Within a total area F_0 there should be n_0 points distributed in such a way that n points are on a certain part F of this area, where $n_0 \gg n$ and $F_0 \gg F$.

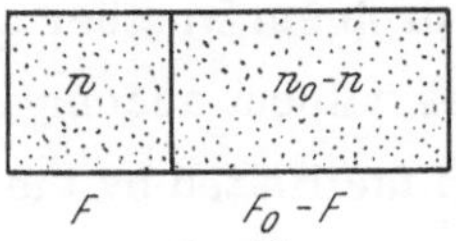

Fig. 19

The probability for this particular distribution has the form

$$w(n) \sim (F/F_0)^n \left(\frac{F_0 - F}{F_0}\right)^{n_0 - n} ,$$

which, modified by a permutation factor due to the indistinctiveness of the points yields the general result

$$w(n) = \frac{n_0!}{n!(n_0 - n)!} \left(\frac{F}{F_0}\right)^n \left(\frac{F_0 - F}{F_0}\right)^{n_0 - n} .$$

In the asymptotic limit $n_0 \to \infty$ the approximation can be used:

$$\frac{n_0!}{(n_0 - n)!} \simeq \frac{n_0^{n_0}}{(n_0 - n)^{n_0 - n}} \simeq \frac{n_0^{n_0}}{n_0^{n_0 - n}} = n_0^n ;$$

and denoting by $\bar{n} = n_0 \frac{F}{F_0}$ the mean number of points on F we obtain for the probability

$$w(n) = \frac{1}{n!} \left(n_0 \frac{F}{F_0}\right)^n \left(1 - \frac{F}{F_0}\right)^{n_0 - n} = \frac{\bar{n}^n}{n!} \left(1 - \frac{\bar{n}}{n_0}\right)^{n_0 - n} .$$

For $n_0 - n \to n_0 \to \infty$ the second factor approximates $\exp(-\bar{n})$, which then finally leads to the Poisson distribution

$$w(n) = \frac{(\bar{n})^n}{n!} e^{-\bar{n}} . \tag{1,4}$$

From this we immediately see that the probability for the emission of a finite number of soft photons vanishes: since $\bar{n} = \infty$ we have $w(n) = 0$ for any finite n.

The discussion of the problem whether a purely elastic scattering of charged particles without accompanying long-wavelength photons is possible can be facilitated by the above result. In classical approximation the scattering process (with associated probability dw_s) is independent of the emission of a number n of soft photons. The total probability for the scattering process plus n soft photons, viz.

$$dw_{(n)} = w(n)\, dw_s ,$$

then vanishes for any finite n $(0, 1, 2, \ldots)$ by virtue of (1,4) and $\bar{n} = \infty$. Summation, however, over all possible values of n yields the real scattering probability as

$$dw = \sum_{n=0}^{\infty} dw_{(n)} = dw_s \sum_{n=0}^{\infty} \frac{\bar{n}^n}{n!} e^{-\bar{n}} = dw_s .$$

This somewhat startling result means that it is not necessary to take into account the interaction of the particle with the infinite number of soft photons explicitly, since the total probability for the scattering process including all soft photons is given by the probability for the basic process; therefore a perturbation expansion is superfluous. This result holds strictly in the limiting case of soft photons (with $\omega \to 0$) only. For any nonvanishing ω, however, the perturbation theory is applicable[+] giving in addition the probability for the emission of hard ($\omega \neq 0$) photons which have to be taken into account the more the larger the uncertainty in the experimental energy-momentum resolution. This point will

+ Here real soft photons lead to a nonvanishing contribution and have to be included, too.

be discussed in more detail in connection with actual calculations.

Finally we want to clarify in this context the question of how much energy soft photons carry with them. We find

$$d\bar{E} = \sum_{n=0}^{\infty} n\hbar\omega\, w(n)\, dw_s =$$

$$= dw_s\, \hbar\omega \sum_{n-1=0}^{\infty} \bar{n}\, e^{-\bar{n}} \frac{\bar{n}^{\,n-1}}{(n-1)!} = \hbar\omega\bar{n}\, dw_s \;;$$

taking into account (1,3), the energy of the infinite number of soft photons is equal to the classical expression for the radiated energy times the probability for the basic process:

$$d\bar{E} = \frac{dI_\omega}{d\omega}\, dw_s \quad .$$

After this discussion of the general principles underlying the classical treatment we examine in the following the validity of the Poisson distribution and develop methods for the actual computation of radiative corrections.

2. Emission of Soft Photons

As mentioned before the emission of soft photons is governed by a Poisson distribution. This statement will be derived in the following, starting from general principles. The probability for the emission of n soft photons by a (classical) current is given by the square of a suitably chosen S-matrix element. The S-matrix can be viewed either in perturbation theory as an infinite sum or one could employ its definition by means of asymptotic field operators. In both cases it is essential that the electromagnetic current is treated as a c-number: we obtain the S-matrix as a simple function of this current and the operators of the electromagnetic field. Also in this treatment, of course, the mean number of emitted photons diverges. It is interesting to note that the perturbation theory here leads to correct results, we shall use, however, the other approach.

First we shall review the formalism. The quantized photon field is described by a vector potential $A^{\mu}(x) \equiv (\Phi, \vec{A})$ where we conveniently use the Coulomb gauge, viz.

$$\Phi = 0, \ \text{div}\, \vec{A} = 0 \ ;$$

with the advantage to deal only with two transverse degrees of freedom for the radiation field. Therefore the plane-wave expansion of $\vec{A}$ has only two terms ($\lambda = 1,2$)

$$\vec{A}(\vec{x}, t) = \int d^3k \sum_{\lambda=1}^{2} \vec{\epsilon}(k, \lambda) A(\vec{k}, \lambda; t)\, e^{i\vec{k}\vec{x}} \ ;$$

where the unit vectors $\vec{\epsilon}(k, \lambda)$ have to be orthogonal to $\vec{k}$ for both values of λ in order to fulfil div $\vec{A} = 0$; in addition we choose them orthogonal to each other:

$$\vec{\epsilon}(k, \lambda) \cdot \vec{\epsilon}(k, \lambda') = \delta_{\lambda\lambda'} \ .$$

Then $\vec{\epsilon}(k, 1)$, $\vec{\epsilon}(k, 2)$, $\vec{k}/|\vec{k}|$ form an orthogonal set. Other important relations are

$$\vec{\epsilon}(-k, 1) = -\vec{\epsilon}(k, 1) \ ,$$

$$\vec{\epsilon}(-k, 2) = \vec{\epsilon}(k, 2) \ ,$$

$$\vec{\epsilon}(k, \lambda)\, \vec{\epsilon}(-k, \lambda') = (-1)^{\lambda}\, \delta_{\lambda\lambda'} \ .$$

Since the free field $\vec{A}_{in}(x)$ is a solution of the wave equation $\Box \vec{A}_{in} = 0$, we can write

$$\vec{A}_{in}(x) =$$

$$= \int \frac{d^3k}{(2\pi)^{3/2}\sqrt{2\omega}} \sum_{\lambda} \vec{\epsilon}(k,\lambda)\,[a_{in}(k,\lambda)\,e^{-i\vec{k}\vec{x}} +$$

$$+ a^{+}_{in}(k,\lambda)\,e^{i\vec{k}\vec{x}}\,] \,. \qquad (1,5)$$

Here a^+, a are creation and annihilation operators respectively of a photon with energy $\omega = |\vec{k}|$ and momentum $\vec{k}$. The well-known commutation relation holds:

$$[a_{in}(k,\lambda),\ a^{+}_{in}(k',\lambda')] = \delta^3(\vec{k}-\vec{k}')\,\delta_{\lambda\lambda'} \,.$$

The scattering process can be represented by a transverse current $\vec{j}(x)$, implying div $\vec{j}(x) = 0$. For our problem we have to solve the field equation $\Box\vec{A}(x) = \vec{j}(x)$ where the current is a c-number in our classical treatment.

By means of the Green's function for the photon field we get as solution of the field equation

$$\vec{A}(x) = \vec{A}_{in}(x) - \int d^4y\, D_{ret}(x-y)\,\vec{j}(y) =$$

$$= \vec{A}_{out}(x) - \int d^4y\, D_{av}(x-y)\,\vec{j}(y) \,.$$

By definition the S-matrix is an operator transforming the "in" states into the "out" states:

$$S^{-1}\vec{A}_{in}(x)\,S = \vec{A}_{out}(x) =$$

$$= \vec{A}_{in}(x) + \int d^4y\,[D_{av}(x-y) - D_{ret}(x-y)]\,\vec{j}(y) =$$

$$= \vec{A}_{in}(x) + \int d^4y\, D(x-y)\,\vec{j}(y) \,. \qquad (1,6)$$

Proceeding to the momentum representation with the integral representations

$$D(z) = D_{av}(z) - D_{ret}(z) =$$

$$= -i \int \frac{d^4k}{(2\pi)^3} e^{-ikz} \delta(k^2) \epsilon(k_0) = -D(-z) \quad ;$$

$$\vec{j}(y) = \int_0^\infty \frac{dk_0}{(2\pi)^{\frac{1}{2}}} \int \frac{d^3k}{(2\pi)^{3/2}} \sum_\lambda \vec{\epsilon}(k,\lambda) [j(k,\lambda) e^{-iky} +$$

$$+ j^*(k,\lambda) e^{iky}]$$

we get

$$\int d^4y\, D(x-y) \vec{j}(y) =$$

$$= \int d^4y\, D(x-y) \int \frac{d^4k'}{(2\pi)^2} \sum_\lambda \vec{\epsilon}(k',\lambda) j(k',\lambda) e^{-ik'y} -$$

$$- \int d^4y\, D(y-x) \int \frac{d^4k'}{(2\pi)^2} \sum_\lambda \vec{\epsilon}(k',\lambda) j^*(k',\lambda) e^{ik'y} =$$

$$= -i \int \frac{d^3k}{(2\pi)\sqrt{2\omega}} \frac{e^{-i\vec{k}\vec{x}}}{\sqrt{2\omega}} \sum_\lambda \vec{\epsilon}(k,\lambda) j(k,\lambda) +$$

$$+ i \int \frac{d^3k}{(2\pi)\sqrt{2\omega}} \frac{e^{i\vec{k}\vec{x}}}{\sqrt{2\omega}} \sum_\lambda \vec{\epsilon}(k,\lambda) j^*(k,\lambda) =$$

$$= \int \frac{d^3k}{(2\pi)^{3/2}\sqrt{2\omega}} \sum_\lambda \vec{\epsilon}(k,\lambda) [-\sqrt{2\pi}\; i \frac{j(k,\lambda)}{\sqrt{2\omega}} e^{-i\vec{k}\vec{x}} +$$

$$+ \sqrt{2\pi}\; i \frac{j^*(k,\lambda)}{\sqrt{2\omega}} e^{i\vec{k}\vec{x}}] \; .$$

Inserting this into (1,5) and (1,6) two equations for a_{in} and a^+_{in} result by comparison of the factors multiplying $e^{-i\vec{k}\vec{x}}$ and $e^{i\vec{k}\vec{x}}$ respectively:

$$S^{-1} a_{in}(k,\lambda) S = a_{in}(k,\lambda) - \\ - \frac{i\sqrt{2\pi}}{\sqrt{2\omega}} j(k,\lambda) = a_{out}(k,\lambda) \;;$$

$$S^{-1} a^+_{in}(k,\lambda) S = a^+_{in}(k,\lambda) + \\ + \frac{i\sqrt{2\pi}}{\sqrt{2\omega}} j^*(k,\lambda) \;. \qquad (1,7)$$

Here the $j(k,\lambda)$ are the coefficients of a representation of $\vec{j}(x)$ analogous to (1,5) and therefore are c-numbers as well. From this we see that the S-operator shifts the "in" fields by a c-number only which facilitates the solution of (1,7) for S by employing an operator identity, applicable if $[A, B]$ is a c-number,

$$e^{-B} A e^{B} = A + [A, B] \;.$$

Therefore S has to be a linear combination of a_{in} and a^+_{in}, which we choose in order to ensure uniformity as

$$S = \exp\{i \int d^3k \sum_\lambda [f(k,\lambda) a^+_{in}(k,\lambda) + \\ + f^*(k,\lambda) a_{in}(k,\lambda)]\} \;.$$

The coefficients $f(k,\lambda)$ are obtained by inserting this expression into (1,7) as

$$f(k, \lambda) = \frac{\sqrt{2\pi}}{\sqrt{2\omega}} j(k, \lambda) \quad .$$

For the computation of matrix elements of this S-matrix

$$S = \exp\left\{ -i \int \frac{\sqrt{2\pi}}{\sqrt{2\omega}} d^3k \sum_\lambda [\, j(k, \lambda)\, a^+_{in}(k, \lambda) + \right.$$

$$\left. + j^*(k, \lambda)\, a_{in}(k, \lambda) \,] \right\} ,$$

we have to rewrite it in terms of normal-ordered operator products, which can be done by virtue of another theorem (again if $[A, B]$ is a c-number):

$$e^{A+B} = e^A e^B e^{-\frac{1}{2}[A, B]} \quad .$$

With this the correct order of operators is ensured and we get

$$S = \exp\left\{ -\pi \int \frac{d^3k}{2\omega} \sum_\lambda |\, j(k, \lambda) \,|^2 \right\} \times$$

$$\times \exp\left\{ -i \int \frac{\sqrt{2\pi}}{\sqrt{2\omega}} d^3k \sum_\lambda [\, j(k, \lambda) a^+_{in}(k, \lambda) \,] \right\} \times$$

$$\times \exp\left\{ -i \int \frac{\sqrt{2\pi}}{\sqrt{2\omega}} d^3k \sum_\lambda [\, j^*(k, \lambda) a_{in}(k, \lambda) \,] \right\} \quad . \tag{1,8}$$

In the following we abbreviate the first factor in (1,8) by A. We are now in a position to evaluate the probability that the current $\vec{j}$ emits n photons with a certain momentum k_i and polarization λ_i. The general expression is

$$P_n(k_i, \lambda_i) = |\langle k_1 \lambda_1 \ldots k_n \lambda_n{}_{in} | S | O_{in} \rangle|^2 \quad .$$

Of more interest, however, is the probability that n photons are emitted in a certain momentum interval K, the polarization being undefined:

$$P_n(K) = \sum_{\lambda_i=1}^{2} \sum_{k_i \epsilon K} |<\ldots k_i \lambda_i \ldots {}_{in} |S| O_{in}>|^2 .$$

The matrix element in question contains these terms of (1,8) only which consist of n creation and no annihilation operators. We therefore get

$$P_n(K) =$$

$$= \sum_{\lambda_i} \sum_{k_i \epsilon K} A^2 |<\ldots k_i \lambda_i \ldots {}_{in} | \frac{(-i)^n}{n!} \{ \int \frac{\sqrt{2\pi}}{\sqrt{2\omega}} d^3k \times$$

$$\times \sum_{\lambda} j(k, \lambda)\, a^+_{in}(k, \lambda) \}^n |O_{in}>|^2 \tag{1,9}$$

Here A^2 obviously has the meaning of the probability that no photon will be emitted, in a rigorous quantum-mechanical treatment it would correspond to the coherent emission and adsorption of virtual photons.

The first summations in (1,9) can be done in the following way: since the k_i are restricted to the interval K, the integration over d^3k is limited to this interval. We then can sum over all <in | states by means of using their completeness relation:

$$P_n(K) = A^2 |<\alpha_{in}| \frac{(-i)^n}{n!} \{ \int_K \frac{\sqrt{2\pi}}{\sqrt{2\omega}} d^3k \sum_{\lambda} j(k, \lambda) a^+_{in}(k, \lambda) \}^n |O_{in}>|^2 =$$

$$= \frac{A^2}{(n!)^2} \langle O_{in} | \{\ldots j^* a \ldots\}^n \cdot 1 \cdot \{\ldots j a^+ \ldots\}^n | O_{in} \rangle =$$

$$= \frac{A^2}{n!} \{ \int_K \frac{2\pi}{2\omega} d^3k \sum_{\lambda=1}^{2} |j(k,\lambda)|^2 \}^n .$$

Here

$$A^2 = \exp \{ - \int \frac{2\pi}{2\omega} d^3k \sum_{\lambda} |j(k,\lambda)|^2 .$$

In the following we shall use the abbreviation

$$\sum_{\lambda} 2\pi \frac{|j(k,\lambda)|^2}{2\omega} = \bar{n}_k ,$$

which represents the mean number of photons with momentum k. The mean number of emitted photons is then given by

$$\bar{n} = 2\pi \int d^3k \frac{\sum_{\lambda} |j(k,\lambda)|^2}{2\omega} = \sum_{\text{all} \vec{k}} \bar{n}_k . \tag{1,10}$$

If we let K be the momentum space as a whole we get for $P_n(K)$ ($K = \text{all} \vec{k}$) indeed a Poisson-distribution:

$$P_n = \frac{e^{-\bar{n}}}{n!} \{\bar{n}\}^n . \tag{1,4}$$

We now introduce in the number of photons subgroups with a distinct value of the momentum and ask for the probability for the emission of n photons, n_k of which have the momentum $\vec{k}$. Because of the relations

$$n = \sum_{\text{all} \vec{k}} n_k ; \quad \bar{n}^n = (\sum_{\text{all} \vec{k}} \bar{n}_k)^n =$$

$$= n! \prod_{\text{all}\,\vec{k}} \frac{\bar{n}_k^{\,n_k}}{n_k!} \;;$$

$$e^{-\bar{n}} = e^{-\Sigma \bar{n}_k} = \prod_{\text{all}\,\vec{k}} e^{-\bar{n}_k} \;;$$

we get

$$P_n(n_k) = \prod_{\text{all}\,\vec{k}} e^{-\bar{n}_k} \frac{(\bar{n}_k)^{n_k}}{n_k!} \quad . \tag{1,11}$$

Here $\bar{n}_k$ represents the mean number of photons, as can be seen easily by:

$$\sum_{n_k} n_k P_{n_k} = \sum_{n_k} n_k e^{-\bar{n}_k} \frac{\bar{n}_k^{\,n_k}}{n_k!} = 1 \cdot \bar{n}_k \quad .$$

The normalization of P_n has been chosen correctly since (1,11) leads to the total probability of one:

$$\sum_{n=0}^{\infty} P_n = 1 \quad .$$

The conclusions following from (1,4) have been discussed in section I.1. However, we want to mention here that the choice of K as the total momentum space is inconsistent with the neglection of the recoil due to photon emission (classical approximation). Despite this fact in actual calculations the distribution (1,4) is used, together, however, with a subsidiary condition which restricts the total momentum attainable by the photons emitted.

This point will be discussed in section 3.

Finally we want to compute explicitly by (1,10) the mean number of photons ($\bar{n}$) under some simplifying assumptions concerning the current $\vec{j}$: An electron scattered in an external field at t = o may have constant velocities $\vec{v}_1$ and $\vec{v}_2$ before and after the scattering, respectively. Then the current is given by

$$\vec{j}(x) = e \lim_{\epsilon \to o} \{\vec{v}_1 \, \delta(\vec{x}-\vec{v}_2 t)\, e^{\epsilon t}\, \Theta(-t) + $$

$$+ \vec{v}_2 \, \delta(\vec{x}-\vec{v}_1 t)\, e^{-\epsilon t}\, \Theta(t) \} \tag{1,12}$$

The first term obviously is defined for $t < o$ only and vanishes for $t \to -\infty$ (by virtue of the factor $e^{\epsilon t}$), it therefore represents the current of the incoming particle. The second term is constructed in an analogous manner. In momentum space we get

$$\vec{j}(k) = \sum_{\lambda} j(k,\lambda)\, \vec{\epsilon}(k,\lambda) = \frac{1}{(2\pi)^2} \int d^4x\, \vec{j}(x)\, e^{ikx} =$$

$$= \frac{e}{(2\pi)^2} \lim_{\epsilon \to o} \{ \vec{v}_1 \int_{-\infty}^{o} dt\, e^{i(\omega - \vec{k}\cdot\vec{v}_1)t + \epsilon t} +$$

$$+ \vec{v}_2 \int_{o}^{\infty} dt\, e^{i(\omega - \vec{k}\cdot\vec{v}_2)t - \epsilon t} \} =$$

$$= - \frac{ie}{(2\pi)^2} \left\{ \frac{\vec{v}_1}{\omega - \vec{k}\cdot\vec{v}_1} - \frac{\vec{v}_2}{\omega - \vec{k}\cdot\vec{v}_2} \right\} ; \tag{1,13}$$

and therefore

$$\sum_{\lambda} | j(k,\lambda)\, \vec{\epsilon}(k,\lambda) |^2 =$$

$$= | j(k,1) |^2 + | j(k,2) |^2 = | j(k) |^2 =$$

$$= \frac{e^2}{(2\pi)^4} \left\{ \frac{\vec{v}_1}{\omega - \vec{k}\cdot\vec{v}_1} - \frac{\vec{v}_2}{\omega - \vec{k}\cdot\vec{v}_2} \right\}^2 ;$$

where we used the fact that the current has transverse components only. The mean number $\bar{n}$ of photons is then given by

$$\bar{n} = \frac{e^2}{(2\pi)^3} \int \frac{d^3k}{2\omega} \left\{ \frac{\vec{v}_1}{\omega - \vec{k}\cdot\vec{v}_1} - \frac{\vec{v}_2}{\omega - \vec{k}\cdot\vec{v}_2} \right\}^2 =$$

$$= \frac{e^2}{(2\pi)^3} \cdot \tfrac{1}{2} \int d\Omega \int \frac{d\omega}{\omega} \left\{ \frac{\vec{v}_1}{1 - \vec{v}_1\cdot\vec{k}_o} - \frac{\vec{v}_2}{1 - \vec{v}_2\cdot\vec{k}_o} \right\}^2 , \qquad (1,14)$$

where $\vec{k}_o = \frac{\vec{k}}{\omega}$.

In the limit $\omega \to o$ the well-known logarithmic divergence shows up causing, however, no principal difficulties but indicating the breakdown of perturbation theory. The form (1,14) for $\bar{n}$ is the relativistic generalization of the expression we derived following equation (1,3):

$$d\bar{n} \sim e^2 \frac{d\omega}{\omega} (\vec{v}_1 - \vec{v}_2)^2 \; .$$

3. Experimental Cross-Section and Radiative Corrections

Until now we have mainly discussed the properties of longwavelength photons accompanying a certain scattering process involving charged particles. In this section, however, we want to analyze in more detail the scattering process itself, especially with regard to the influence of the accompanying photon field on experimental measurements.

One always has to keep in mind that any moving charged particle is surrounded by an undefined number of soft photons, therefore the definition of a cross-section without these quanta has no sense, theoretically and experimentally it is zero. In each experimental setup there are inherent uncertainties: particles cannot be measured to have exactly this energy and are scattered under that angle, but one necessarily has to allow for finite energy and angle (= momentum) intervals in order to get finite counting rates. It is therefore impossible to improve the accuracy of the measurements beyond a certain limit. This fact has to be taken into account in theoretical calculations of cross-sections: the experimental setup with its uncertainties must be known in order to get comparable results. This experimentally predefined energy-momentum-uncertainty exactly corresponds to the energy and momentum which are carried away undetected by the accompanying photons. We say undetected because these photons are not observed, their number is unknown and they are not treated as individual particles in the calculation. They constitute a collective effect describable by the classical probability theory, whereas applying the perturbation theory without caution would lead to difficulties since then these photons are treated individually, which is improper in case of long-wavelength ($\omega = o$) photons.

The occurrence of soft photons is connected intimately with the basic uncertainties of the measurements; in the following we shall cast this fact into a suitable mathematical form. For a relatively simple process we want to outline the method: as the basic process we take electron-proton scattering ($e + p \to e' + p'$) for which the perturbation theory in zeroth order yields a doubly differential cross-section $d^2\sigma(\Omega)$, with Ω as the solid angle of the observed particles (electrons). The foregoing discussion, however, implies that the physically relevant cross-section should include the accompanying electromagnetic radiation as well and therefore should be of the form $d^6\sigma(\Omega, k)$, where the four-vector k represents energy and momentum of this field. The magnitude of the energy carried away by the field depends on the specific experiment. We define a function $\rho(k)$ as the probability that in a specific experiment a four-momentum loss of magnitude k is undetected. This function $\rho(k)$ then restricts the total possible spectrum to the contributions essential for the experiment in question, and we get for the experimental cross-section

$$d^2\sigma_{exp}(\Omega) = \int d^4k\ \rho(k)\frac{d^6\sigma(\Omega,k)}{d^4k} .$$

As long as the momentum loss remains small the emission process is independent of the basic scattering process and we can factorize the cross-section in the form

$$d^6\sigma(\Omega,k) = d^4P(k)\cdot N\cdot d^2\sigma_0(\Omega) . \qquad (1,15)$$

Here $d^4P(k)$ is the probability for the momentum loss to lie in d^4k and can be calculated by means of classical statistical mechanics under the assumption of sufficiently small momentum losses which leads to the Poisson distribution discussed in the foregoing section. The fact that photons with finite energy ω have also to be taken into account is represented by the normalization factor N. Their influence on the cross-section can be calculated by means of the perturbation theory. This intermixing of classical and perturbation theoretical treatment will be discussed later, we only want to anticipate here that by properly fixing the limit up to which photons can be regarded as soft the quantity N differs only little from one. This constitutes a certain arbitrariness of the method which, however, is justified by its success. Therefore the experimentally observable cross-section in this approximation is

$$d^2\sigma_{exp}(\Omega) = \int d^4P(k)\ \rho(k)\ N\ d^2\sigma_0(\Omega) . \qquad (1,16)$$

To calculate explicitly the probability $d^4P(k)$ we use the Poisson distribution (1,11) as the probability for the occurrence of n photons, n_{k_i} of which having momentum k_i (these momenta can be assumed to take discrete values corresponding to a division of phase space into sufficiently small cells). Furthermore we have the constraint that the total momentum loss does not exceed k; this leads to the form

$$d^4P(k) = \sum_{\text{all } n_{k'}} P_{n_{k'}}(n_{k'})\, \delta^4\Big(\sum_{\vec{k}'} k' n_{k'} - k\Big)\, d^4k \ .$$

After some rearrangements and with the help of the Fourier-representation of the δ-function we get the relation

$$d^4P(k) = \sum_{n_{k'}} \Big\{ \prod_{\vec{k}'} e^{-\bar{n}_{k'}} \frac{(\bar{n}_k)^{n_{k'}}}{n_{k'}!} \Big\} \int \frac{d^4x}{(2\pi)^4} e^{-ikx} e^{ix \sum_{\vec{k}'} k' n_{k'}}\, d^4k =$$

$$= \prod_{\vec{k}'} \sum_{n_{k'}} \int \frac{d^4x}{(2\pi)^4} \frac{(\bar{n}_{k'} e^{ixk'})^{n_{k'}}}{n_{k'}!} e^{-\bar{n}_{k'}} e^{-ikx}\, d^4k =$$

$$= \prod_{\vec{k}'} \int \frac{d^4x}{(2\pi)^4} e^{-\bar{n}_{k'}(1 - e^{ik'x})} e^{-ikx}\, d^4k \ ;$$

$$d^4P(k) = \Big\{ \int \frac{d^4x}{(2\pi)^4} e^{-h(x) - ikx} \Big\}\, d^4k \ ; \tag{1,17}$$

where we introduced the function h (x) as

$$h(x) = \sum_{\vec{k}'} \bar{n}_{k'} (1 - e^{ik'x}) \ . \tag{1,18}$$

In principle the summation over momenta in the δ-function and therefore in (1,18) should include only values for k' which are small enough to justify the neglect of the recoil of the particles due to the emitted photons. Otherwise the factorization (1,15) and the use of the Poisson distribution is improper. It has turned out, however, that the summation can be extended far beyond the limiting energy ϵ of soft photons without causing large errors. It is even suitable to take the summation up to an energy E characteristic for the specific scattering process (e.g. the energy of the scattered electron). According to (1,1) and (1,2) ϵ should be small compared to E and of order of magnitude of the electron's mass. In the following we are going to justify the fact that we can replace ϵ by E as upper limit in the summation over $\vec{k}'$. But it should again be noted that we thereby introduce a certain arbitrariness which reflects itself in the limited exactness of the results. Therefore the probability $d^4P(k)$ which we obtain in this way is restricted to small momentum losses, or in other words to sharp experimental resolution.

In order to explicitly compute (1,17) we need the mean number of photons with momentum $\vec{k}$ ($\bar{n}_k$) in an analytic form. Because of the definition (1,10) and the special result (1,14) we expect also in general the form

$$d\bar{n}(\omega) = \int_\Omega d^3k\, \bar{n}(k) = \int_\Omega 2\pi k^2 dk d\Omega \frac{\sum_{\lambda=1}^{2} |j(k,\lambda)|^2}{2\omega} =$$

$$= \beta \frac{d\omega}{\omega} . \tag{1,19}$$

The angular integration is contained in the factor β, and we canceled k^2 in the numerator against an ω^{-2} resulting from $|j(k,\lambda)|^2$. How to calculate explicitly β in certain cases will be shown later in detail, the result being that β is a number very small compared to one, defined by the kinematics of the process in question. We assume therefore β to be known and discuss in the following the resulting form of $d^4P(k)$.

For the simplest case of an experiment with a certain energy uncertainty the integration over d^3k in (1,17) can be performed in a trivial way resulting in an energy-distribution function $dP(\omega)$:

$$dP(\omega) = \int_{\text{momenta}} d^4P(k) =$$

$$= \frac{1}{(2\pi)^4} \int dt\, d\omega \int d^3x \int d^3k\, e^{-i\vec{k}\vec{x}}\, e^{-h(\vec{x},t)} e^{i\omega t} =$$

$$= \frac{d\omega}{2\pi} \int_{-\infty}^{+\infty} dt\, e^{-h(O,t)} e^{i\omega t} \; .$$

By replacing the summation over k' in (1,18) by an integral and with the help of (1,19) we get

$$h(0,t) = \beta \int_0^E \frac{d\omega'}{\omega'} (1 - e^{-i\omega' t}) \; .$$

It is interesting to note that this integral is no longer divergent at the lower limit (for soft photons), therefore the infinite number of soft photons gives a finite contribution. The remaining integrations in $dP(\omega)$

$$dP(\omega) = \frac{d\omega}{2\pi} \int dt \left\{ \exp\left\{-\beta \int_0^E \frac{d\omega'}{\omega'} (1 - e^{-i\omega' t})\right\} e^{i\omega t} \right\}$$

can be executed in general form for $\omega < E$. The somewhat lengthy calculation is done in section II.4, here we shall only quote the result and continue to discuss the physical consequences:

$$dP(\omega) = \beta \frac{d\omega}{\omega} \left(\frac{\omega}{E}\right)^{\beta} \cdot \frac{e^{-\beta \ln\gamma}}{\Gamma(1+\beta)} \; , \tag{1,20}$$

with $\gamma = 1.781$ (Euler's constant). The series expansion for small β of the last factor in (1,20)

$$\frac{e^{-\beta \ln \gamma}}{\Gamma(1+\beta)} = 1 - \frac{\pi^2}{12}\,\beta^2 + \ldots ,$$

shows that we can safely replace it by one due to the smallness of β. For $\omega = 0$ also in (1,20) no divergence occurs by virtue of the factor ω^{β}. The insertion of (1,20) into (1,16) yields together with the assumed energy resolution ΔE ($\rho(k) = 1$ for $\omega < \Delta E$ and zero otherwise)

$$d^2\sigma_{exp} = \left(\int_0^{\Delta E} dP(\omega) \right) N\, d^2\sigma_0 = \left(\frac{\Delta E}{E}\right)^{\beta} N\, d^2\sigma =$$

$$= e^{\beta \ln \frac{\Delta E}{E}} N\, d^2\sigma_0 \;. \tag{1,21}$$

In order to compare this form with the result of the perturbation theory we expand the exponential function to get

$$d^2\sigma_{exp} \simeq \left(1 + \beta \ln \frac{\Delta E}{E}\right) N\, d^2\sigma_0 \;. \tag{1,21'}$$

The well-known formula by Schwinger for the cross-section of electron scattering in an external field, again with an energy resolution ΔE, in perturbation approximation is given by

$$\left(\alpha = \frac{e^2}{4\pi}\right)$$

$$d^2\sigma_s(\Delta E) = d^2\sigma_0 \left\{ 1 - \frac{4\alpha}{\pi}\left[\left(\ln \frac{E}{\Delta E} - \frac{13}{12}\right) \tfrac{1}{2} \left(\ln \frac{|q^2|}{m^2} - 1\right) + \right.\right.$$

$$\left.\left. + \frac{17}{72} + \ldots \right]\right\} \;.$$

Provided that ΔE is small compared to E the main contribution stems from the term $\ln E/\Delta E$ and we see that the factor

$$\frac{2\alpha}{\pi}\left(\ln \frac{|q^2|}{m^2} - 1\right)$$

corresponds to β in (1,21') as we shall show later explicitly. Therefore we can rewrite the Schwinger result as

$$d^2\sigma_s(\Delta E) = d^2\sigma_0 \left\{ 1 + \beta \ln \frac{\Delta E}{E} + \text{smaller terms} \right\} . \qquad (1,22)$$

Since β itself is small we can factor out $(1 + \beta \ln \frac{\Delta E}{E})$ and arrive at a form analogous to (1,21')

$$d^2\sigma_s(\Delta E) = (1 + \beta \ln \frac{\Delta E}{E}) \left[1 + \text{small terms} \right] d^2\sigma_0 \ . \qquad (1,22')$$

Now we are able to calculate the so far undetermined factor N in perturbation approximation.
The general form for N is

$$N = 1 + \alpha\gamma^{(1)} + \alpha^2\gamma^{(2)} + \alpha^3 \gamma^{(3)} + \ldots ,$$

where we only know the first term linear in α of this perturbation expansion hoping that we can neglect terms of higher order.

As a result we have found that the approximation (1,21') is to be replaced by (1,21) which represents the contribution of soft photons in a more general form. This replacement also corresponds to an intuitive suggestion by Schwinger to consider $d^2\sigma_0$ $(1 - \delta)$ a series expansion to first order of the exact formula $d^2\sigma_0 \exp(-\delta)$, the last expression leading to more accurate results for the energy resolution becoming extremely sharp. As an example, for $\Delta E = 0$ we get from (1,21) the exact result $d^2\sigma_{exp} \sim e^{-\infty} = 0$ whereas the approximation (1,21') would predict an infinitely large cross-section in the same way as the result of a first order perturbation expansion.

Expression (1,21) therefore represents a recipe for practical calculations of an experimentally observable cross-section: One extracts from the result of perturbation calculations the contribution of soft photons and takes this expression as argument of the exponential function; this procedure corresponds to a summation over all soft quanta. The remaining contributions of the perturbation result are then incorporated in factor N. It should be noted that the above expression (1,21) was obtained for a special case only: electron scattering in an external field; the experimental resolution being such that any energy loss larger than ΔE will be detected for sure ($\rho(k) \equiv 0$, except for $\omega \leq \Delta E$); no angular resolution. However, the essential features of the theory of radiative corrections with special consideration of soft photons showed up clearly in the discussion of this special case. In order to describe realistic scattering experiments successfully we have to generalize our arguments; this involves no principal difficulties but the mathematical apparatus will become more complex.

As a first step we have to find a general form for the probability function $d^4P(k)$. The result, which we only quote here, is of extremely simple structure:

$$d^4P(k) = \frac{e^{-\beta \ln \gamma}}{\Gamma(1+\beta)} \beta \frac{d\omega}{\omega} \left(\frac{\omega}{E}\right)^{\beta} A(\vec{u})\, d^3u \; ; \qquad (1,23)$$

with

$$\vec{u} = \vec{k}/\omega \; .$$

The new additional factor $A(\vec{u})$ represents a many-photon-distribution function, normalized to one

$$\int A(\vec{u})\, d^3u = 1 \; . \qquad (1,24)$$

This function depends on the one-photon-distribution function $f(\vec{n})$ in a mathematically complicated manner; in practical calculations, however, one may equate the two functions

$$A(\vec{u}) = f(\vec{n}) \; ,$$

without committing large errors or may approximate $A(\vec{u})$ in another manner. The main point is that the normalization condition (1,24) must be fulfilled.

Furthermore, a more general description of the experimental conditions involves a function ρ characterizing the energy and momentum resolution. The step function used in the foregoing is only a crude approximation of reality, a Gaussian distribution would be more appropriate. We therefore assume

$$\rho(k) = \exp\{-\frac{\omega^2}{2(\Delta E)^2}\} \exp\{-\sum_1^3 a_{ij} \frac{k_i k_j}{2(\Delta p)^2}\} . \qquad (1,25)$$

Here ΔE and Δp are the maximal energy and momentum uncertainties, respectively. The experimentally determined matrix of coefficients a_{ij} has matrix elements of maximal magnitude one.

If we now insert (1,23) and (1,25) into (1,16) we get an expression valid for a large number of various experiments:

$$d^2\sigma_{exp}(\Omega) = C(\rho)\, N\, d^2\sigma_0(\Omega) .$$

We took advantage of the separability condition (1,15) which implies that N and $d^2\sigma_0$ are independent of k. The factor $C(\rho)$ contains the integration over d^4k for soft photons

$$C(\rho) = \int d^4P(k)\, \rho(k) =$$

$$= \beta \int \frac{d\omega}{\omega} \left(\frac{\omega}{E}\right)^{\beta} \exp\{-\frac{\omega^2}{2(\Delta E)^2}\}$$

$$\int A(\vec{u}) \exp\{-\Sigma a_{ij} \frac{k_i k_j}{2(\Delta p)^2}\} d^3u . \qquad (1,26)$$

It is convenient to combine $C(\rho)$ and N in the form

$$C(\rho)\, N = 1 - \delta .$$

Here δ, the "radiative correction", usually is a quantity between 10 % and 30 %. This is the value by which the experimental cross-section differs from the theoretical one in first order perturbation approximation. Through $C(\rho)$, soft photons contribute most in case of very sharp experimental resolution. So far we have not explicitly calculated β, this will be done in the next section. Also some examples for the actual computation of radiative corrections will be discussed there.

Finally we want to stress the following: The integration over $d\omega'$ in (1,18) has been extended to a characteristic energy E of the process in question, which is large compared to the limiting energy ϵ of soft photons. This same energy also shows up in Schwinger's formula (1,22) and even in the same combination $\ln \frac{\Delta E}{E}$. It is therefore not only convenient to choose E as upper limit of integration but it can also be shown that this choice is permissible. Integration in (1,18) up to ϵ changes (1,21) into

$$d^2\sigma_{exp} = \exp\left\{ \beta \ln\left(\frac{\Delta E}{\epsilon}\right)\right\} N\, d^2\sigma_0 \ .$$

Since the result should be independent of the arbitrarily chosen cut-off ϵ the factor N has to contain a term of the form $\ln \frac{\epsilon}{E}$ in order to cancel ϵ. Indeed, this is the case. This interplay of perturbation theory and classical Poisson-distribution will be more intelligible after a thorough discussion of the perturbation theory itself. This is the subject of chapters II, III.

4. Calculations of Radiative Corrections

In the first place we shall explicitly compute the parameter β introduced in (1,19). It obviously represents one of the essential quantities of the theory and is defined by an integral over the photon angles as

$$\beta = \pi \int d\Omega\, \omega^2 \sum_{\lambda=1}^{2} |j(k,\lambda)|^2 . \tag{1,27}$$

The form (1,13) of the electron current can be generalized without great difficulties; by inspecting the calculation leading from (1,12) to (1,13) we see that the annihilation of an electron corresponds to a current

$$\vec{j}(k) = -\frac{ie}{(2\pi)^2} \frac{\vec{v}}{\omega - \vec{k}\cdot\vec{v}} = \frac{i}{(2\pi)^2} \frac{|e|\vec{v}}{\omega - \vec{k}\cdot\vec{v}} .$$

For the case of creation the sign is changed, also in case of a positively charged particle. Z_i denoting the sign of the charge of the ith particle and by means of the sign-function ϵ_i

$$\epsilon_i = \begin{cases} +1 & \text{for creation} \\ -1 & \text{for annihilation} \end{cases}$$

the generalization of (1,13) reads

$$\vec{j}(k) = \sum_{\lambda} j(k,\lambda)\, \vec{\epsilon}(k,\lambda) =$$

$$= \frac{i|e|}{(2\pi)^2} \sum_{\substack{\text{all charged} \\ \text{particles}}} \frac{\epsilon_i Z_i \vec{v}_i}{\omega - \vec{k}\cdot\vec{v}_i} ;$$

or in the relativistic notation

$$j^{\mu}(k) = \frac{i|e|}{(2\pi)^2} \sum_i \frac{\epsilon_i Z_i p_i^{\mu}}{(p_i, k)} , \tag{1,28}$$

where

$$p_i^{\mu} = (E, \vec{p}_i) = E(1, \vec{v}_i) \text{ and } k^{\mu} = (\omega, \vec{k}) = \omega(1, \vec{n}) .$$

By virtue of the denominator $(\omega - \vec{k}\cdot\vec{v})$ the current has a maximum in case of relativistic particles ($|\vec{v}| \to 1$) if $\vec{k}$ and $\vec{v}$ are parallel, i.e. photons are emitted preferentially in the direction of the current. In contrast to this result the condition of transversality (since $\bar{n} \sim \sum_{\lambda} |j(k,\lambda)|^2 = |j(k,1)|^2 + |j(k,2)|^2 = |j_{\perp}(k)|^2$) favors directions of $\vec{k}$ normal to the current. The resulting angular distribution out of these two competing conditions requires a more detailed investigation for any specific process.

Since we now know the current $\vec{j}$ explicitly the integral in (1,27) can be evaluated. As a simple example we consider an electron-positron colliding-beam experiment as shown in Fig. 20.

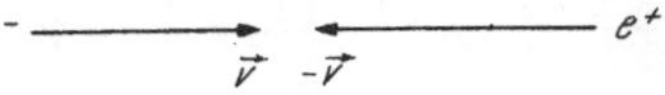

Fig. 20

The transverse part of the current has the form

$$\vec{j}_{\perp}(k) = \frac{i|e|}{(2\pi)^2}\left\{\frac{(+\vec{v})}{\omega - \vec{k}\cdot\vec{v}} - \frac{(-\vec{v})}{\omega + \vec{k}\cdot\vec{v}}\right\}\sin\Theta =$$

$$= \frac{i|e|\vec{v}}{(2\pi)^2\,\omega}\;\frac{2\sin\Theta}{1 - v^2\cos^2\Theta}$$

where we used the notation $\vec{k}\cdot\vec{v} = \omega\, v\cos\Theta$. Here we shall not discuss the current corresponding to the outgoing particles. The integral

$$\beta = \frac{1}{2}\int \frac{d\varphi\, d(\cos\Theta)}{(2\pi)^3}\,\omega^2\,\frac{4e^2v^2}{\omega^2}\,\frac{\sin^2\Theta}{(1-v^2\cos^2\Theta)^2} =$$

$$= \frac{2e^2v^2}{(2\pi)^2}\int_{-1}^{+1}\frac{(1-x^2)\,dx}{(1-v^2x^2)^2}$$

can be evaluated by elementary methods:

$$\int_{-1}^{+1} dx \; \frac{1-x^2}{(1-v^2x^2)^2} =$$

$$= \frac{1}{2v^2} \left\{ \frac{v^2+1}{v} \ln \frac{1+v}{1-v} - 2 \right\} .$$

With $e^2/4\pi = \alpha$ we then get

$$\beta = \frac{\alpha}{\pi} \left\{ \frac{1+v^2}{v} \ln \frac{1+v}{1-v} - 2 \right\} .$$

In the extreme-relativistic limit $(v \to 1)$ we can rewrite this expression by means of $E^2(1-v^2) = m^2$ as

$$\beta_{ER} = \lim_{v \to 1} \frac{\alpha}{\pi} \left\{ \frac{v^2+1}{v} \ln \frac{E^2(1+v)^2}{m^2} - 2 \right\} =$$

$$= \frac{2\alpha}{\pi} \left\{ \ln \left(\frac{2E}{m}\right)^2 - 1 \right\} .$$

In kinematically more complicated cases where incoming and outgoing particles have to be described simultaneously, or if the particle momenta are not coplanar, the above discussed angular integration turns out to be quite elaborate. It is then more favorable to use the covariant notation where (1,27) can be evaluated in a general form by means of the expression (1,28) for the current.

Due to the vanishing divergence of the current we have $|j_\perp|^2 = |j_\mu|^2$; β is therefore given as

$$\beta = \frac{\omega^2 e^2}{2(2\pi)^2} \int d\Omega \sum_{(i,j)} \frac{\epsilon_i \epsilon_j Z_i Z_j (p_i, p_j)}{(p_i, k)(p_j, k)} =$$

$$= \frac{\alpha \omega^2}{(2\pi)^2} \sum_{(i,j)} \epsilon_i \epsilon_j Z_i Z_j (p_i, p_j) \int \frac{d\Omega}{(p_i, k)(p_j, k)} .$$

We evaluate this integral by means of the Feynman method:

$$\int \frac{d\varphi d(\cos\Theta)}{(k,p_i)(k,p_j)} = \int_0^1 dz \int \frac{d\varphi d(\cos\Theta)}{(k,p_z)^2} ;$$

where $p_z = p_i z + p_j (1-z)$,

$$\int \frac{d\varphi d(\cos\Theta)}{(k,p_z)^2} = \frac{2\pi}{\omega^2} \int_{-1}^{+1} \frac{d(\cos\Theta_z)}{(E_z - P_z\cos\Theta_z)^2} =$$

$$= \frac{4\pi}{\omega^2} \frac{1}{p_z^2} .$$

The terms with i = j are most easily calculated:

$$\int_0^1 \frac{dz}{p_z^2} = \int_0^1 \frac{dz}{p_i^2} = \frac{1}{p_i^2} .$$

A somewhat more elaborate calculation yields for the terms with $i \neq j$:

$$\int_0^1 \frac{dz}{p_z^2} = \frac{1}{(p_i,p_j)} \frac{1}{Q_{ij}} \ln \frac{1+Q_{ij}}{1-Q_{ij}} ;$$

where

$$Q_{ij} = \sqrt{1 - (p_i^2 p_j^2)/(p_i, p_j)^2} .$$

Taken altogether we get for β the expression

$$\beta = -\frac{\alpha}{\pi} \{ \sum_i z_i^2 +$$

$$+ \sum_{i>j} Z_i Z_j \epsilon_i \epsilon_j \frac{1}{Q_{ij}} \ln \frac{1+Q_{ij}}{1-Q_{ij}} \} . \qquad (1,29)$$

To check this result we apply (1,29) to the previously calculated $e^- - e^+$ colliding beam experiment. We have

$$p_1 = (E, O, O, P) ; p_1^2 = p_2^2 = m^2 = E^2 (1 - v^2) ; Q_{12} = \frac{2v}{1+v^2}$$

$$p_2 = (E, O, O, - P) ; (p_1, p_2) = E^2 + P^2 = E^2 (1+v^2) .$$

Insertion of this into (1,29) gives

$$\beta = -\frac{\alpha}{\pi} \{ 2 - \frac{1+v^2}{2v} \ln \frac{(1+v)^2}{(1-v)^2} \} =$$

$$= \frac{\alpha}{\pi} \{ \frac{1+v^2}{v} \ln \frac{1+v}{1-v} - 2 \} ,$$

in conformity with the result previously obtained.

At this point we want to mention one essential detail. The conformity of the results of the covariant method and the intuitive one, which we discussed earlier, strongly depends on the transversality condition $\operatorname{div} \vec{j} = O$. If, for instance, we neglect in an electron-electron colliding beam experiment the current due to the outgoing particles the two methods lead to different results since then $\operatorname{div} \vec{j} \neq O$, which implies $|j_\perp|^2 \neq |j_\mu|^2$. We are justified in neglecting the outgoing current, however, assuming that the masses of the particles created (e.g. protons, antiprotons) are large, their contribution to $\beta \sim \ln \frac{2E}{M}$ being therefore much smaller than the corresponding electron contributions.

An estimate on the order of magnitude of β can be obtained from the extreme-relativistic expression. For the case of an $e^+ - e^-$ experiment with beam energies of 1000 MeV we get

$$\beta = \frac{2\alpha}{\pi} \{ 2 \ln \frac{2E}{m} - 1 \} \cong 0.07 .$$

This number is characteristic for experiments of this type; β has been called "Bond" factor because of its significant numerical value.

As a further example we shall discuss relativistic ($E^2 \gg m^2$) electron-proton scattering. The kinematics are chosen as (indices 1 and 3 relate to the electron, 2 and 4 to the proton):

$$p_1 = (E_1, P_1, O, O) ; p_2 = (M, 0, 0, 0) ; p_{3,4} =$$

$$= (E_{3,4}, P_{3,4} \cos \Theta_{3,4}) \; P_{3,4} \sin \Theta_{3,4}, 0) .$$

Then

$$Q_{ee} = \sqrt{1 - \frac{m^4}{(p_1, p_3)^2}} \simeq 1 - \frac{2m^4}{q^4} ;$$

with $q^2 = (p_1 - p_3)^2 = 2m^2 - 2(p_1, p_3) \simeq - 2(p_1, p_3)$.

Here we introduced explicitly the momentum transfer q as a significant quantity, in our case $|\vec{q}\,|$ is equal to P_4, the momentum of the outgoing proton. We also have

$$Q_{pp} = \sqrt{1 - \frac{M^2}{E_4^2}} = \frac{P_4}{E_4} ;$$

and since $(p_1, p_2) = (p_3, p_4)$ and $(p_1, p_4) = (p_2, p_3)$

$$Q_{12} = Q_{34} = Q_1 = \sqrt{1 - m^2 / E_1^2} \simeq 1 - m^2 / 2E_1^2 ,$$

and

$$Q_{14} = Q_{23} = Q_3 = \sqrt{1 - m^2 / E_3^2} \simeq 1 - m^2 / 2E_3^2 .$$

The Bond factor then consists of the following terms:

$$\beta_{ee} = -\frac{\alpha}{\pi}\{2 + \ln\frac{m^4}{q^4}\} = \frac{2\alpha}{\pi}\{\ln(\frac{-q^2}{m^2}) - 1\} ;$$

$$\beta_{pp} = -\frac{\alpha}{\pi}\{2 - \frac{E_4}{P_4}\ln\frac{1+P_4/E_4}{1-P_4/E_4}\} \simeq \frac{2\alpha}{\pi}\frac{|\vec{q}|^2}{3M^2} ;$$

$$\beta_{ep} = -\frac{\alpha}{\pi}\{\frac{2}{Q_1}\ln\frac{1+Q_1}{1-Q_1} - \frac{2}{Q_3}\ln\frac{1+Q_3}{1-Q_3}\} \simeq$$

$$\simeq \frac{2\alpha}{\pi} 2\ln\frac{E_1}{E_3} ;$$

which, taken altogether, yield

$$\beta = \frac{2\alpha}{\pi}\{\ln(\frac{-q^2}{m^2}) - 1 + \frac{1}{3}\frac{|\vec{q}|^2}{M^2} + 2\ln\frac{E_1}{E_3}\} .$$

Without the last two terms, often called dynamic corrections, this is exactly the same result as quoted earlier. From the following numerical example it may be seen that the contribution of the dynamic corrections are but small:

$$E_1 = 2000 \text{ MeV};\ E_3 = 1000 \text{ MeV};\ \Theta_3 = 60^\circ ;$$

$$\beta = \frac{1}{215}\{15.9 - 1 + 1.1 + 1.4\} = 0.076 .$$

From these calculations it is obvious that the Bond factor can be determined without great difficulties. But this is not the whole problem, we still have to evaluate the correction factor $C(\rho)$ which contains the experimental details in a crucial way.

Equation (1,26) defines $C(\rho)$ where the experimental energy and momentum resolutions are incorporated by ΔE and Δp respectively (together with the coefficient matrix a_{ij} determining the directions of momenta). With the abbreviations

$$x^2 = \frac{\omega^2}{2(\Delta p)^2} \quad \text{and } y = \frac{\Delta p}{\Delta E}$$

we have $C(\rho)$ in the form

$$C(\rho) = \beta \left(\frac{\sqrt{2}\,\Delta p}{E}\right)^{\beta} \int_0^{E/\sqrt{2}\Delta p} \frac{dx}{x} x^{\beta} e^{-x^2 y^2} \int d^3u\, A(\vec{u})\, e^{-a_{ij} u_i u_j x^2} \tag{1,30}$$

Due to the strong convergence of the x-integration we can extend the upper limit to infinity without committing a large numerical error. This then allows for an analytic evaluation of the integral.

From the definition of Γ-functions follows

$$\int_0^\infty dx\, x^{\beta-1} e^{-ax^2} = \tfrac{1}{2} a^{-\beta/2}\, \Gamma(\beta/2)\,, \tag{1,31}$$

a relation which will be frequently used.

Now we shall briefly discuss two characteristic limiting cases of (1,30) : with $\Delta E = \Delta p \to \infty$ $(x \to 0,\ y \to 1)$, $C(\rho)$ tends to one and therefore $d^2\sigma_{exp} = N\, d^2\sigma_0$; for ext~emely sharp resolution, $\Delta E = \Delta p = O$ $(x \to \infty)$, the integral (1,30) vanishes, so does $d^2\sigma_{exp}$ as we had expected it to do.

The case of energy resolution alone $(\Delta p = \infty)$ has been discussed in the foregoing section, for the case of momentum resolution alone $(\Delta E = \infty)$ we expect from (1,30) a result proportional to $(\Delta p)^{\beta} \sim (1 + \beta \ln \Delta p)$, depending only on Δp.

The remaining discussion of the general case we conduct by means of the two typical examples for which we have calculated β already.

We assume that the experimental setup of the colliding beam experiment consists of two counters, measuring only the two components of the momenta normal to the primary beam with a certain

finite resolution, as shown in Fig. 21

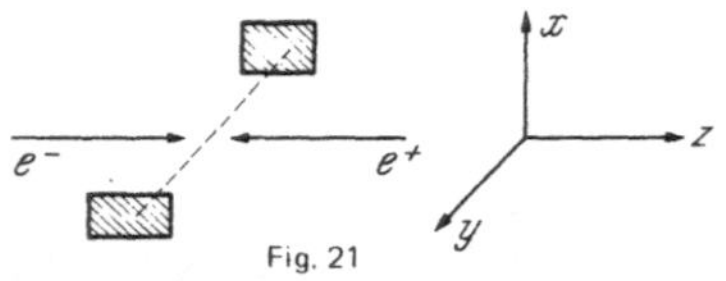

Fig. 21

$\Delta E = \infty \ldots y = 0$;

$\Delta p_x = \Delta p_z = \Delta p \ldots a_{xx} = a_{zz} = 1,$

all other $a_{ij} = 0 \ldots \Delta p_y = \infty$.

Then (1,30) takes the form

$$C = \beta \left(\frac{\sqrt{2}\,\Delta p}{E} \right)^{\beta} \int_0^{\infty} dx\, x^{\beta - 1} \int d^3u\, A(\vec{u})\, e^{-(u_x^2 + u_z^2)x^2} .$$

We have already mentioned that the exact form of the many-photon distribution $A(\vec{u})$ is not crucial; this will be confirmed by the following discussion. A simple and appropriate assumption would be that photons are emitted in direction of the $(e^+ - e^-)$ beam only; this fact can be represented by two δ-functions. In addition, we incorporate an isotropic contribution leading to

$$A(\vec{u}) = \frac{b}{2} \{ \delta(u_z - 1) + \delta(u_z + 1) \} \delta(u_x)\, \delta(u_y) +$$

$$+ \frac{1-b}{4\pi} \delta(|\vec{u}| - 1) .$$

Obviously $A(\vec{u})$ is normalized to 1. The constant b serves as a free parameter adjustable according to the relative contributions of the δ - functions and the isotropic part. The integrations are then trivial.

$$I(x) = \int d^3u\, A(\vec{u})\, e^{-(u_x^2 + u_y^2)x^2} =$$

$$= b\, e^{-x^2} + \frac{1-b}{2} \int_{-1}^{+1} d\xi\, e^{-(1-\xi^2)x^2} .$$

With (1,31) we get

$$C = \left(\frac{\sqrt{2}\,\Delta p}{E}\right)^{\beta} \left\{ \int_0^{\infty} dx\, x^{\beta-1}\, I(x) \right\} =$$

$$= \left(\frac{\sqrt{2}\,\Delta p}{E}\right)^{\beta} \beta \left\{ \frac{b}{2}\, \Gamma\left(\frac{\beta}{2}\right) + \right.$$

$$\left. + \int_{-1}^{+1} d\xi\, (1-\xi^2)^{-\beta/2}\, \frac{1-b}{4}\, \Gamma\left(\frac{\beta}{2}\right) \right\} .$$

The remaining integral can again be expressed in terms of Γ-functions as

$$\int_{-1}^{+1} d\xi\, (1-\xi^2)^{-\beta/2} = \int_0^{\pi} (\sin\Theta)^{1-\beta}\, d\Theta =$$

$$= \frac{\pi\, \Gamma(1-\beta)\, 2^{\beta}}{\Gamma\left(\frac{3}{2}-\frac{\beta}{2}\right) \Gamma\left(\frac{1}{2}-\frac{\beta}{2}\right)} .$$

After some rearrangements we finally get the result

$$C = \left(\frac{\sqrt{2}\,\Delta p}{E}\right)^{\beta} \Gamma\left(1+\frac{\beta}{2}\right) \left\{ b + \right.$$

$$\left. + (1-b)\, \frac{2^{-\beta} \left[\Gamma\left(1-\frac{\beta}{2}\right)\right]^2}{(1-\beta)\, \Gamma(1-\beta)} \right\} \simeq$$

$$\simeq \left(\frac{\sqrt{2}\,\Delta p}{E}\right)^{\beta} \Gamma\left(1+\frac{\beta}{2}\right) \left\{ b + \right.$$

$$\left. + (1-b)\left[1+\beta(1-\ln 2) \ldots \right] \right\} , \qquad (1,32)$$

having expanded the last term for small β. From this we see that there is no big difference whether we assume a totally anisotropic ($b = 1$) or a totally spherical symmetric ($b = o$) distribution $A(\vec{u})$. For a $\beta = 0.07$, for instance, the difference is $\beta(1-\ln 2) \simeq 2\,\%$ only. The fact that these two contradicting limiting cases lead to almost identical results lends plausibility to the conjecture that the exact form of $A(\vec{u})$ does not enter significantly. It is not surprising therefore that the choice

$$A(\vec{u}) = f(\vec{n}) = \frac{\alpha v^2}{\beta\pi^2} \frac{\sin^2\Theta}{(1-v^2\cos^2\Theta)^2}$$

gives a result numerically very similar to the one above. Here $f(\vec{n})$ is the one-photon distribution function obtainable from $|j_\perp(k)|^2$ with the requirement to be normalized to one, resulting in the additional factors in the expression above. With this we have to evaluate the integral

$$\beta \int_0^\infty dx\, x^{\beta-1} \int d(\cos\Theta)\, d\varphi\, f(\Theta)\, e^{-x^2(1-\sin^2\Theta\sin^2\varphi)} =$$

$$= \int d(\cos\Theta)\, d\varphi \frac{\alpha v^2}{\beta\pi^2} \frac{\sin^2\Theta}{(1-v^2\cos^2\Theta)^2} (1-\sin^2\Theta\sin^2\varphi)^{-\beta/2}\Gamma(1+\frac{\beta}{2}).$$

In order to facilitate the computation we expand $(\dots)^{-\beta/2}$ in a power series; to first order we then get

$$2\pi \int \frac{1-\cos^2\Theta}{(1-v^2\cos^2\Theta)^2} d(\cos\Theta) +$$

$$+\frac{\beta}{2} \int \frac{\sin^4\Theta\sin^2\varphi}{(1-v^2\cos^2\Theta)^2} d(\cos\Theta)\, d\varphi\ .$$

The first term is known already from the evaluation of the Bond factor by means of (1,27); applying the same method we calculate the second term

$$\frac{\beta}{2}\int \ldots = \pi\frac{\beta}{2}\int_{-1}^{+1}\frac{(1-x^2)^2\,dx}{(1-v^2x^2)^2},$$

which in the limit $v \to 1$ gives $\beta\pi$. The correction factor C in the relativistic limit then reads

$$C = \Gamma(1+\frac{\beta}{2})\left(\frac{\sqrt{2}\,\Delta p}{E}\right)^{\beta}\{1+\frac{\alpha}{\pi}\}, \qquad (1,32')$$

where terms of higher order in β have been neglected. This result (1,32') conforms very well with (1,32).
In order to get an order of magnitude for C we again assume an experiment with beam energies of 1000 MeV, $\beta = 0.07$ and $\Delta p =$ $= 10$ MeV; (1,32') then gives

$$C = \Gamma(1.0035)\,(0.0141)^{0.07}\,(1.002) \simeq 0.73.$$

Since the factor N, resulting from the contributions of the hard photons, differs not much from unity we therefore expect the radiative correction to be

$$\delta = 27\ \% \quad .$$

With $\Delta p = 100$ MeV the contribution of radiative correction decreases to 15 %. These values show the orders of magnitude to be expected for radiative corrections in high energy experiments, together with the fact that sharper momentum (or energy) resolution corresponds to higher values for δ.

For our following example, the calculation of C for an electron-proton scattering experiment, we shall first describe the experimental setup in more detail. An electron of high energy is scattered elastically by a proton at rest and will be observed, its final momentum making a certain angle with the incident direction. The energy of the scattered electron is measured too, where we can assume that the relevant uncertainty in measuring the energy (ΔE) is the dominant quantity, the uncertainties in the determination of the angle and primary energy being negligible compared with ΔE. This ΔE, however, is not identical with the one used in (1,26), it rather represents a combination of ΔE, Δp and a_{ij} as defined there. The value of ΔE, which we are using here, represents the uncertainty of the electron's energy and therefore has a critical influence on the energy and momentum (≙ angular) spectrum of the photons emitted. It will then be our first task to relate this ΔE to the maximal energy of soft photons ω_{max} (which of course is different for different directions). Having found this relation it is no longer necessary to convert ΔE into the quantities occurring in (1,26) but rather to use a form of (1,26) adapted for this case

$$C(\rho) = \beta \int_0^E \frac{d\omega}{\omega} \left(\frac{\omega}{E}\right)^\beta \int d\varphi \; d(\cos\Theta) f(\Theta, \varphi) \times$$

$$\times \exp\left\{-\frac{\omega^2}{2\omega_{max}^2}\right\}. \qquad (1,33)$$

This expression can be evaluated analytically, at least in some approximation. By means of Fig. 22 we introduce our notation.

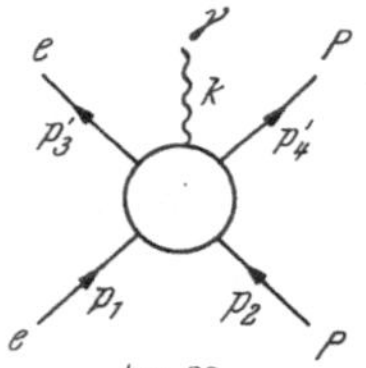

Fig. 22

Assuming that a real photon is emitted with four-momentum k we can relate the experimental uncertainty in energy ΔE to the maximum possible photon energy. If we denote by p_i' the respective momenta in case of the additional photon being emitted and by p_i otherwise, energy and momentum conservation yields

$$p_4^2 = M^2 = (p_1 + p_2 - p_3)^2 ,$$

$$p_4'^2 = M^2 = (p_1 + p_2 - p_3' - k)^2 \ .$$

By equating these two expressions we get

$$(p_3 - p_3')(p_1 + p_2) = (p_1 + p_2 - p_3')k = (p_4 + p_3 - p_3')k \ . \tag{1,34}$$

If the scattering angle is known more precisely than the energy E_3 we can put

$$E_3 - E_3' = \Delta E, \ \Delta P = \vec{p}_3 - \vec{p}_3' \ ;$$

$$\Theta_{13} = \Theta_{13}' \ .$$

For relativistic electrons ($v \to 1$) we have $\Delta E = \Delta P$, and (1,34) passes on to

$$\Delta E(E_1 + M - E_1 \cos\Theta_{13}) =$$

$$= \omega(E_4 + \Delta E - P_4 \cos\Theta_{\gamma 4} - \Delta E \cos\Theta_{\gamma 3}) \ ,$$

which gives

$$\omega_{max} = \frac{\Delta E \cdot M[1 + \frac{E_1}{M}(1 - \cos\Theta_{13})]}{E_4(1 - v_4 \cos\Theta_{\gamma 4}) + \Delta E(1 - \cos\Theta_{\gamma 3})} =$$

$$= \frac{\Delta E}{E_3} \frac{M E_1}{(1 - v_4 \cos\Theta_{\gamma 4}) + \Delta E(1 - \cos\Theta_{\gamma 3})} \ .$$

Here we have used the kinematic relation

$$E_1 = \eta E_3 \ ; \ \eta = 1 + \frac{E_1}{M}(1 - \cos\Theta_{13}) \ .$$

Except in case of $v_4 \simeq 1$, i.e. the recoil proton being an extremely relativistic particle, we may neglect the second term in the denominator by virtue of $\Delta E \ll E_4$ to get

$$\omega_{max} = \frac{\Delta E}{E_3} \frac{ME_1}{E_4} \frac{1}{1-v_4 \cos\Theta_{\gamma 4}} . \tag{1,35}$$

This is the equation of an ellipsoid of revolution, prolate in the direction of $\vec{p}_4$. In the limit $v_4 \to 0$, i.e. in case of a nonrelativistic recoil proton, the ellipsoid degenerates into a sphere and ω_{max} is independent of the proton's direction. Some interesting cases for the maximal photon energy in various directions are listed below:

a) photon in direction of $\vec{p}_1$: $\omega_1 = \eta^2 \Delta E = \frac{E_1^2}{E_3^2} \Delta E$;

since $\cos\Theta_{14} = \frac{P_4}{E_1} \frac{E_1+M}{E_4+M}$, and

$$\frac{ME_1(E_4+M)}{E_4E_1(E_4+M)-P_4^2(E_1+M)} = \frac{ME_1(E_4+M)}{E_1M^2+E_1E_4M-E_4^2M+M^3} =$$

$$= \frac{E_1(E_4+M)}{E_1(E_4+M)-(E_4+M)(E_4-M)} = \frac{E_1}{E_3} .$$

b) photon in direction of $\vec{p}_3$: $\omega_3 = \Delta E$;

since $\cos\Theta_{34} = \cos(\Theta_{13}+\Theta_{14}) = \frac{E_1^2-E_3^2-P_4^2}{2EP}$, and

$$\frac{2E_3P_4}{E_4 \cdot 2E_3P_4 - P_4(E_1^2-E_3^2-P_4^2)} =$$

$$= \frac{2E_3}{2E_3E_4-(E_4-M)(E_1+E_3)+(E_4+M)(E_4-M)} =$$

$$= \frac{2E_3}{2E_3E_4+(E_4-M)2(M-E_3)} =$$

$$= \frac{E_3}{E_4M-M^2+E_3M} = \frac{E_3}{E_1M} .$$

c) photon in direction of $\vec{p}_4$: $\omega_4 = \frac{\Delta E}{E_3} \frac{ME_1}{E_4 \pm P_4}$;

which follows immediately from $\cos \Theta_{\gamma 4} = \pm 1$.

It should be mentioned that (1,35) is valid also in case $v_4 \to 1$. The approximation used seems to be inappropriate only if at the same time the photon is emitted in direction of the outgoing proton. In this case, however, $\Theta_{\gamma 3} = \Theta_{13} + \Theta_{14}$ and

$$\frac{\Delta E_3}{E_4}(1 - \cos \Theta_{\gamma 3}) = \frac{\Delta E_3 ME_1}{E_3E_4^2} .$$

We then get

$$\omega_{max} = \frac{\Delta E}{E_3} \frac{ME_1}{E_4} \frac{1}{1-v_4 \cos \Theta_{\gamma 4} + (\Delta E_3 ME_1)/(E_3E_4^2)}$$

Therefore, even in case of $v_4 \to 1$ and the photon direction coinciding with the proton direction, the last term in the denominator still is negligible and (1,35) is valid also in this case.

The maximum energy of the photon as given by (1,35) we then insert into (1,33); since $\omega_{max} << E$ we again replace

$\int_0^E d\omega$ by $\int_0^\infty d\omega$ to get

$$C(\rho) = \beta \int_0^\infty d\omega\, \omega^{\beta-1} E^{-\beta} \int d(\cos\Theta)\, d\varphi f(\Theta,\varphi) \exp\left\{\frac{-\omega^2}{2\omega_{max}^2(\Theta,\varphi)}\right\} =$$

$$= \frac{\Gamma(1+\frac{\beta}{2})}{\left(\frac{E}{\sqrt{2}}\right)^\beta} \int d(\cos\Theta)\, d\varphi f(\Theta,\varphi) \{\omega_{max}(\Theta,\varphi)\}^\beta .$$

As we have seen from our first example the exact form of the photon distribution is immaterial; for reasons of convenience we therefore assume that the photons are emitted solely in the direction of the relativistic electrons:

$$f(\Theta,\varphi) = \tfrac{1}{2}\,\delta\left(\vec{n} - \frac{\vec{p}_1}{P_1}\right) + \tfrac{1}{2}\,\delta\left(\vec{n} - \frac{\vec{p}_3}{P_3}\right) = f_1 + f_2 .$$

The angular integration is trivial, leading to two terms. Here we experience the arbitrariness in the choice of E. It seems most convenient to put $E = E_1$ in the first integral (containing f_1) and $E = E_3$ in the second one. Finally we arrive at

$$C(\rho) = 2^{\beta/2}\, \Gamma(1+\frac{\beta}{2}) \left\{ \tfrac{1}{2}\left(\frac{\eta^2 \Delta E}{E_1}\right)^\beta + \tfrac{1}{2}\left(\frac{\Delta E}{E_3}\right)^\beta \right\} , \qquad (1,36)$$

where we have used the results of cases a) and b) for ω_{max}. Again

$$\beta \cong \frac{2\alpha}{\pi} \ln\left(\frac{|q^2|}{m^2}\right)$$

is a small quantity (of order of magnitude 0.08 for the example under consideration), therefore an expansion up to terms linear in β seems sufficient:

$$C(\rho) \cong 1 - \delta = 1 - C\frac{\beta}{2} + (\ln 2)\frac{\beta}{2} +$$

$$+\frac{\beta}{2}\{\ln(\frac{E_1}{E_3}\frac{\Delta E}{E_3}) + \ln\frac{\Delta E}{E_3}\} + \ldots \qquad (1,36')$$

To get a representative number out of this we evaluate the term in curly brackets only with $E_1 = 2000$ MeV, $\eta = 2$, $\Delta E/E = 1\ \%$, $\Theta_{13} = 60^{\circ}$, and have

$$\delta \cong -0.04\{\ln 2 + 2\ln\ 0.01\} = +0.34\ .$$

For such a sharp ΔE, however, it is more appropriate to employ (1,36): This gives $\delta \simeq 30\ \%$. For $\frac{\Delta E}{E_3} = 10\ \%$ then (1,36') is sufficiently accurate to give

$$\delta \simeq 16\ \%\ .$$

Also these numbers are of expected magnitude. Here we do not want to improve the method; chapter II will include a more detailed discussion of this process.

In this chapter, by means of purely classical methods, we computed the correction factor $C(\rho)$ and the magnitude of radiative corrections assuming that the probability of emission of hard photons introduces no significant differences ($N \simeq 1$). From the derivation of the results we see that they are the more accurate the sharper the relevant energy-momentum resolution, since then the contribution of hard photons decreases whereas the contribution of soft photons increases due to the typical $\ln\frac{\Delta E}{E}$ - term. In high-energy scattering experiments equipped with high resolution power radiative corrections are of sizable amount; these can be determined, however, by rather simple calculations with relatively high accuracy. In the framework of QED also a quite successful formalism has been developed to compute radiative

corrections to extreme accuracy. Unfortunately the mathematical apparatus involved is quite complicated. The development of this formalism together with some applications for typical examples will be the subject of chapters III and IV.
The following chapter II represents so to speak an intermediate step between classical and quantumelectrodynamical treatment and contains some calculations of principal importance.

II. Summation over Soft-Photon Contributions

This chapter represents an intermediate step between classical theory and quantumelectrodynamical treatment. Soft and hard photons will be separated in a rigorous way; the contribution of soft photons (classical current) as a whole can be taken into account as in the previous chapter (here we shall employ the derivation by means of the perturbation expansion of the S-matrix), the application of the rules of QED to the hard-photon contribution introduces no principal difficulties (infrared divergences).

1. Separation of Soft and Hard Photons

The vector potential representing the electromagnetic field is split up (in the interaction representation) into two contributions:

1. $a_\mu^{(s)}$ for $\omega \leq \epsilon$ ... soft photons interacting with a classical current $J_\mu(x)$;

2. $a_\mu^{(h)}$ for $\omega > \epsilon$ ... hard photons, where the usual Feynman rules are to be applied.

Field operators of these different types of particles commute. The choice of the limiting energy $\epsilon \neq 0$ is arbitrary as long as conditions (1,1) and (1,2) are fulfilled, guaranteeing the validity of the classical behavior in the first case.

Then the Hamiltonian representing the interaction of photons with a charged particle consists of two additive terms

$$H(x_0) = H^J(x_0) + H^{\Psi}(x_0) \ .$$

Here

$$H^J(x_0) =$$

$$= \int\limits_{\omega \leq \epsilon} \frac{d^3k}{(2\pi)^3 \sqrt{2\omega}} \left[a_{\mu}^{(s)}(\vec{k})\, e^{i\omega x_0} + \right.$$

$$\left. + a_{\mu}^{(s)*}(-\vec{k})\, e^{i\omega x_0} \right] \int d^3x\, J^{\mu}(x)\, e^{i\vec{k}\cdot\vec{x}} \tag{2,1}$$

represents the interaction of soft photons. For hard photons we get an analogous expression (with $\omega > \epsilon$) with the corresponding photon operator $a_{\mu}^{(h)}$ and the quantumelectrodynamical current (e.g. $e\bar{\Psi}\gamma_{\mu}\Psi$). Therefore H^{Ψ} contains explicitly the field operators of the charged particles, quantum field theory then is appropriate for this contribution.

Since we treat soft and hard photons as essentially different particles and H^J contains no Ψ- operators two important relations hold:

$$[H^J(x_0), H^{\Psi}(x_0')] = 0 \ ,$$

$$[H^J(x_0), H^J(x_0')] = \text{c-number} \ . \tag{2,2}$$

With these the following statements can be made

A. The S-matrix can be factorized: $S = S^J \cdot S^{\Psi}$

B. The evaluation of S^J involves no principal difficulties.

Statement A can be derived easily. The S-matrix is considered the sum over all contributions to arbitrary order in perturbation theory:

$$S = \sum_{n=o}^{\infty} S^{(n)} ,$$

$$S^{(n)} = \frac{(-i)^n}{n!} \int_{-\infty}^{+\infty} dx_1^o \dots \int_{-\infty}^{+\infty} dx_n^o \, T\{H^J(x_1^o) +$$

$$+ H^{\Psi}(x_1^o)) \dots (H^J(x_n^o) + H^{\Psi}(x_n^o))\} =$$

$$= \frac{(-i)^n}{n!} \int_{-\infty}^{+\infty} \dots \int_{-\infty}^{+\infty} dx_n^o \sum_{k=o}^{n} \binom{n}{k} T\{H^J(x_1^o) \dots$$

$$\dots H^J(x_k^o)\} T\{H^{\Psi}(x_{k+1}^o) \dots H^{\Psi}(x_n^o)\} ,$$

where T { ... } is the well-known time ordering operator. The last line results by virtue of (2,2). The S-matrix as a whole can be factorized in the same way as each term in the series:

$$S = \sum_{n=o}^{\infty} \sum_{k=o}^{n} (-i)^k / k! \, \frac{(-i)^{n-k}}{(n-k)!} \int_{-\infty}^{+\infty} \dots$$

$$\dots \int dx_1^o \dots dx_k^o \dots dx_n^o \, T\{H^J(x_1^o) \dots$$

$$\dots H^J(x_k^o)\} T\{H^{\Psi}(x_{k+1}^o) \dots H^{\Psi}(x_n^o)\} =$$

$$= S^J \cdot S^{\Psi}$$

where

$$S^{J/\Psi} = \sum_{n=0}^{\infty} \frac{(-i)^n}{n!} \int_{-\infty}^{+\infty} \dots \int dx_1^0 \dots dx_n^0 \, T\{ H^{J/\Psi}(x_1^0) \dots$$

$$\dots H^{J/\Psi}(x_n^0) \}, \text{ q.e.d.}$$

Statement B involves a somewhat lengthy calculation. As a first step we evaluate the second commutator in (2,2) (omitting in the course of the calculation the indices J and s):

$$C(x_0, x'_0) \equiv [H(x_0), H(x'_0)] =$$

$$= \int \frac{d^3k}{(2\pi)^3 \sqrt{2\omega}} \int \frac{d^3k'}{(2\pi)^3 \sqrt{2\omega'}} \int d^3x \, e^{i\vec{k}\cdot\vec{x}} \times$$

$$\times \int d^3x' \, e^{i\vec{k}'\vec{x}'} \int \frac{d^4p}{(2\pi)^2} e^{-ipx} J^\mu(p) \times$$

$$\times \int \frac{d^4q}{(2\pi)^2} e^{-iqx'} J^\nu(q) \{ [a_\mu(\vec{k}),$$

$$a_\nu^*(-\vec{k}')] e^{-i\omega x_0 + i\omega' x'_0} +$$

$$+ [a_\mu^*(-\vec{k}), a_\nu(\vec{k}')] e^{i\omega x_0 - i\omega' x'_0} \} .$$

With

$$[a_\mu(\vec{k}), a_\nu^*(-\vec{k}')] = - g_{\mu\nu} \, \delta^3(\vec{k} + \vec{k}')$$

we get

$$C(x_0, x'_0) = - \frac{1}{(2\pi)^3} \int \frac{d^3k}{2\omega} \int \frac{dp_0}{\sqrt{2\pi}} \int \frac{dq_0}{\sqrt{2\pi}} \int d^3p \int d^3q \times$$

$$\times \int \frac{d^3x}{(2\pi)^3}\, e^{i(\vec{k}+\vec{p})x} \int \frac{d^3x'}{(2\pi)^3}\, e^{-i(\vec{k}-\vec{q})\vec{x}'} \times$$

$$\times\ J^{\mu}(\vec{p},p_o)\, J_{\mu}(\vec{q},q_o)\, e^{-ip_o x_o}\, e^{-q_o x'_o} \times$$

$$\times \{ e^{-i\omega(x_o-x'_o)} - e^{i\omega(x_o-x'_o)} \} =$$

$$= -\frac{1}{(2\pi)^3} \int \frac{d^3k}{2\omega} \int \frac{dp_o}{\sqrt{2\pi}} \int \frac{dq_o}{\sqrt{2\pi}} \times$$

$$\times\ J_{\mu}(-\vec{k},p_o)\, J^{\mu}(\vec{k},q_o)\, e^{-i(p_o x_o + q_o x'_o)} \{\ldots\}.$$

By means of the Hamilton operator and the S-matrix we shall now describe the evolution with time of states in general and thereby obtain a relation between these quantities.

The equation of motion of a state Ψ has the form:

$$i\dot{\Psi} = H\Psi .$$

The transformed state $\Phi = \exp\{i\Sigma(t)\}\Psi$ satisfies the equation

$$i\dot{\Phi} = H'\Phi ,$$

with

$$H' = \exp\{i\Sigma\}(H - i\tfrac{\partial}{\partial t})\exp\{-i\Sigma\} .$$

This new operator can be expanded in a series as

$$H' = (1 + i\Sigma - \tfrac{1}{2}\Sigma\cdot\Sigma\ldots)\{H(1 - i\Sigma - \tfrac{1}{2}\Sigma\cdot\Sigma\ldots) -$$

$$- i[\, i\dot{\Sigma} - \tfrac{1}{2}(\dot{\Sigma}\Sigma + \Sigma\dot{\Sigma}) +$$

$$+ \frac{i}{6}(\dot{\Sigma}\Sigma\Sigma + \Sigma\dot{\Sigma}\Sigma + \Sigma\Sigma\dot{\Sigma}) \ldots]\} =$$

$$= H + i [\Sigma, H] + \tfrac{1}{2} [[\Sigma, H], \Sigma] +$$

$$+ \dots - \dot{\Sigma} - \frac{i}{2} [\Sigma, \dot{\Sigma}] + \frac{1}{6} [[\dot{\Sigma}, \Sigma] \Sigma] + \dots .$$

By means of the special choice $\Sigma(x_0) = \int_{-\infty}^{x_0} H(t)\,dt$ we get $\dot{\Sigma}(x_0) = H(x_0)$ and $\Sigma(-\infty) = 0$. Then the commutators $[\Sigma, H]$ and $[\dot{\Sigma}, \Sigma]$ are c-numbers and the series for H' terminates after a few terms:

$$H' = H - \dot{\Sigma} + i \{ [\Sigma, H] + \tfrac{1}{2} [\dot{\Sigma}, \Sigma] \} =$$

$$= \frac{i}{2} \int_{-\infty}^{x_0} dx'_0 \, [H(x'_0), H(x_0)] \; ;$$

or

$$H'(x_0) = - \frac{i}{2} \int_{-\infty}^{x_0} dx'_0 \; C(x_0, x'_0) \; .$$

The time evolution of the state Φ can be described by means of the S-matrix. Since

$$\Phi(x'_0) = V(x'_0, x_0) \, \Phi(x_0) \; ,$$

where

$$V(x'_0, x_0) = \exp \{ i \int_{x_0}^{x_0} H'(\tau)\, d\tau \} \; ,$$

we get by rewriting the expression in terms of Ψ and letting x_0 to $-\infty$, x'_0 go to $+\infty$:

$$e^{i\Sigma(x'_0)} \Psi(x'_0) = V(x'_0, x_0) \, e^{i\Sigma(x_0)} \Psi(x_0)$$

or

$$\Psi(\infty) = S \ \Psi(-\infty)$$

with

$$S = e^{-i\Sigma(\infty)} V(\infty, -\infty) e^{i\Sigma(-\infty)} =$$

$$= e^{-i\Sigma(\infty)} e^{-i \int_{-\infty}^{\infty} H'(\tau) d\tau} =$$

$$= e^{-i \int_{-\infty}^{+\infty} H(\tau) d\tau} e^{-\frac{1}{2} \int_{-\infty}^{+\infty} dx_0 \int_{-\infty}^{x_0} dx_0' C(x_0, x_0')} \equiv$$

$$\equiv e^{-i\Sigma} e^{i\Theta} .$$

With this expression statement B is essentially demonstrated. The expression $\Sigma = \int_{-\infty}^{+\infty} H^J (x_0) dx_0$ has to be evaluated with the help of (2,1) and (1,28), which is easily done

$$\Sigma = \frac{1}{2\pi} \int \frac{d^3k}{\sqrt{2\omega}} \{ J^{\mu *}(\vec{k}) a_\mu^{(s)} (\vec{k}) +$$

$$+ J^\mu(\vec{k}) a_\mu^{(s)*} (\vec{k}) \} . \tag{2,3}$$

The second factor requires no detailed investigation since as a pure phase factor it constitutes no observable effect. It has to be shown that Θ is a real quantity:

$$\Theta = -\frac{i}{2} \int_{-\infty}^{+\infty} dx_0 \int_{-\infty}^{x_0} dx_0' C(x_0, x_0') =$$

$$= \frac{i}{2} \frac{1}{(2\pi)^3} \int \frac{d^3k}{2\omega} \int_{-\infty}^{+\infty} dx_0 \int \frac{dp_0}{\sqrt{2\pi}} \frac{dq_0}{\sqrt{2\pi}} J^\mu J_\mu \times$$

$$\times \left\{ \frac{e^{-i(p_o+q_o)x_o}}{i(\omega-q_o)} + \frac{e^{-i(p_o+q_o)x_o}}{i(\omega+q_o)} \right\}$$

$$\Theta = -\tfrac{1}{2} \frac{1}{(2\pi)^3} \int \frac{d^3k}{2\omega} \int dq_o\, J^{\mu}(\vec{k}, q_o)\, J^{*}_{\mu}(\vec{k}, q_o) \left\{ \frac{1}{\omega - q_o} + \frac{1}{\omega + q_o} \right\},$$

which is real indeed.

The statements

$A : S = S^J \cdot S^{\Psi}$

$B : S^J = e^{i\Theta} e^{-i\Sigma}$

will be of much importance in the following discussions. The consequence of A is that matrix elements of soft and hard photons can be treated separately, the first of which can then be evaluated easily in general by means of B and the known classical current J_{μ}. The next sections will therefore deal with those different matrix elements.

2. Virtual Soft Photons

As general initial state we take one containing a certain number of electrons (or other particles) and possibly some hard photons:

$$|\Phi) = |p_1, p_2, \ldots p_r; \gamma_h) \ .$$

At first no real soft photons are to be allowed. The corresponding final state then is

$$(\Psi| = (p_1', \ldots p_r'; \gamma_h'| \ .$$

The matrix element for the process in question $(\Psi|S|\Phi)$ implicitly also contains transitions involving virtual photons. These photons, however, constitute no difficulties in our formalism since for hard photons no logarithmic divergences occur due to the fact that the integral $\int \frac{d\omega}{\omega}$ has a lower limit $\epsilon \neq 0$. By virtue of the coupling with a classical current the contributions of soft photons to each order can be summed up without the occurrence of a divergence.

Since neither $(\Psi|$ nor $|\Phi)$ contain soft photons the transition matrix element factorizes to

$$(\Psi|S|\Phi) = (\Psi|S^J\, S^{\Psi}|\Phi) = (\Psi|S^{\Psi}|\Phi)\,(0|S^J|O)\ . \tag{2,4}$$

Here the operator S^J acts in the Hilbert space of soft photons; since here for both initial and final state only the vacuum occurs the second factor represents the basic process in question, may be under the influence of hard quanta, and has to be evaluated by means of QED.

At first we investigate the effect of those virtual soft photons. Therefore we divide the momentum space into small cells of magnitude Δ_r, accordingly we have a discrete manifold of momenta k_r $(r = 1,2\ \ldots)$ in the place of continuous vectors $\vec{k}$. Due to this division some important relations are changed which we briefly list in the following:

$$\int \frac{d^3k}{(2\pi)^{3/2}}\, f(\vec{k}) \rightarrow \sum_{r=1}^{\infty} \Delta_r\, f(k_r)\ ;$$

$$[\, a^{(\lambda)}(k_r),\ a^{*(\lambda')}(k_s)\,] = (2\pi)^{3/2}\, \delta_{\lambda\lambda'}\, \frac{\delta_{rs}}{\Delta_r}\ ;$$

$$\Sigma \rightarrow \sum_r \frac{\Delta_r}{\sqrt{2\omega_r}}\, \sqrt{2\pi}\, \{ J^{\mu *}(k_r)\, a_{\mu}^{(s)}(k_r) +$$

$$+ J^{\mu}(k_r)\, a_{\mu}^{(s)*}(k_r) \}\ ;$$

$$a_\mu(k_r) = \epsilon_\mu^{(\lambda)} a^{(\lambda)}(k_r) .$$

For the evaluation of $(O|S^J|O)$ we use statement B together with (2,3): $(O|S^J|O) = e^{i\Theta}(O|e^{-i\Sigma}|O)$ and expand the second exponential function in the relevant series. The term Σ^{21} we write down explicitly

$$\Sigma^{21} = \sum_{r_1, r_2, \dots r_{21}} (2\pi)^l \frac{\Delta r_1 \dots \Delta r_{21}}{\sqrt{2\omega_{r_1} \dots 2\omega_{r_{21}}}} \{J^* a^{(s)} +$$

$$+ J a^{(s)*}\}_1 \dots \{J^* a^{(s)} + J a^{(s)*}\}_{21} .$$

This expression has to be time ordered with respect to the operators a leading to a sum of products of these operators together with all possible contractions. Since the matrix element is taken between vacuum states applied to which the operators a give zero only terms remain where all operators are contracted. Altogether we have $(2l)!/2^l l!$ such possibilities:

By virtue of

$$[a_\mu(k_r), a^*_\nu(k_s)] = -(2\pi)^{3/2} g_{\mu\nu} \frac{\delta_{rs}}{\Delta r} ,$$

the contraction of $a_{\mu_i}(k_i)$ with $a^*_{\mu_j}(k_j)$ yields a factor

$$\rho = -\sum_i \frac{\Delta_i}{2\omega_i} J^*_\mu J^\mu (2\pi)^{5/2} =$$

$$= \sum_i \frac{\Delta_i}{2\omega_i} |\vec{J}(k_i)|^2 (2\pi)^{5/2} \to 2\pi \int_{\omega \le \epsilon} \frac{d^3k}{2\omega} |\vec{J}(\vec{k})|^2 . \tag{2,5}$$

The matrix element in question therefore is

$$(O|S^J|O) = e^{i\Theta} \sum_{l=0}^{\infty} \frac{(-i)^{2l}}{(2l)!} (O|\Sigma^{2l}|O) =$$

$$= e^{i\Theta} \sum_{l=0}^{\infty} \frac{(-1)^l}{2^l l!} \rho^l (O|O) = e^{i\Theta} e^{-\rho/2} \; . \tag{2,6}$$

Now we may finally omit the phase factor $e^{i\Theta}$ since it cancels with its complex conjugate in the calculation of transition probabilities. The second term we already know in a form similar to (2,5)-compare (2,5) with (1,8) and (1,19). From this we see

$$\rho = \pi \int d\Omega \, |\vec{J}(\vec{k})|^2 \omega^2 \int_0^{\epsilon} \frac{d\omega}{\omega} = \beta \int_0^{\epsilon} \frac{d\omega}{\omega} \to +\infty \; ,$$

which shows that this quantity obviously diverges. This corresponds to the fact that an infinite number of virtual soft photons is involved since ρ represents the mean number of virtual quanta. Therefore the elastic S-matrix element in question and, according to (2,4), also the total transition amplitude vanish:

$$(\Psi|S|\Phi) = (\Psi|S^{\Psi}|\Phi) \, e^{-\frac{1}{2}\rho} \underset{\rho\to\infty}{\to} O \; .$$

This result we already are familiar with: a purely elastic process without real soft photons has a vanishing cross-section.

In this connection we want to point out one exception: if both $|\Phi)$ and $(\Psi|$ contain only uncharged particles the current identically vanishes and therefore also $\rho \equiv O$. We then have $(\Psi|S|\Phi) = (\Psi|S^{\Psi}|\Phi)$, but such processes (e.g. photon-photon-scattering) do not interest us here. In order to obtain finite cross-sections we therefore have to include in our description also real soft photons which are emitted during the scattering process.

3. Real Soft Photons

We now want to consider cross-sections proportional to

$$\sum_n |(\Psi + n\gamma^s|S|\Phi)|^2$$

where the total energy of the emitted soft photons is limited to a finite value ΔE by experimental uncertainties in observation. There should exist n_i soft photons with (discrete) momenta k_i and polarization λ_i, their total number being $n : \sum_i n_i = n$. In our matrix element

$$(\Psi, n_1, n_2 \ldots |S|\Phi) = (\Psi|S^\Psi|\Phi)\,(n_1, n_2 \ldots |S|O) \,, \qquad (2,7)$$

the second factor again can be evaluated explicitly. Therefore we represent the state containing the real soft photons by the successive application of field operators to the vacuum:

$$(n_1, n_2 \ldots | = (O| \prod_{r=1}^{\infty} \frac{1}{\sqrt{n_r!}} \{a^{(\lambda_1)}(k_1)\}^{n_1} \{a^{(\lambda_2)}(k_2)\}^{n_2} \ldots$$

The nth term in the series expansion of the S^J-matrix element has the form

$$(n_1, n_2 \ldots | \frac{(-i)^n}{n!} \Sigma^n |O) =$$

$$= (O| \prod_{r=1}^{\infty} \frac{1}{\sqrt{n_r!}} \{a_{(1)}^{n_1} a_{(2)}^{n_2} \ldots\}$$

$$\frac{(-i)^n}{n!} \sum_{s_1 \ldots s_n} \frac{\Delta_{s_1} \ldots \Delta_{s_n}}{\sqrt{2\omega_{s_1} \ldots 2\omega_{s_n}}} (2\pi)^{n/2} \times$$

$$\times [J^* a(k_{s_1}) + J a^*(k_{s_1})] \dots [J^* a(k_{s_n}) +$$

$$+ J a^*(k_{s_n})] \,|0) .$$

In this case only terms without contractions remain, i.e. only those which come about after the commutation of the n a-operators of the final state with the n a^*-operators of Σ^n. The above expression is therefore

$$\frac{(-i)^n}{n!} \prod_{r=1}^{\infty} \frac{1}{\sqrt{n_r!}} \{ (2\pi)^{5/4} \sqrt{\Delta_r / 2\omega_r} \quad \times$$

$$\times \; J^{\mu}(k_r) \, e_{\mu}^{(\lambda_r)} \}^{n_r} \, (0|0) \; n$$

Here the last factor n arises from the various possibilities of commuting a and a*. This term now would correspond to a process without any additional virtual quanta. In general we therefore have to include also terms of order $m > n$.

If $m = n + 2l$, again l additional contractions are possible which can be done in $(n + 2l)! \,/ (2^l l! n!)$ number of ways. This is the number of possibilities to pick l ordered pairs out of $n + 2l$ elements. From the previous section, we already know the result of such a contraction, we therefore get ($\sum_i n_i = n$, $m = n + 2l$):

$$(n_1, n_2 \dots |S^J|O) = \sum_{m=0}^{\infty} \frac{(-i)^m}{m!} (n_1, n_2 \dots |\Sigma^m|O) =$$

$$= \sum_{m=0}^{n-1} 0 + \frac{(-i)^n}{n!} (n_1, n_2 \dots |\Sigma^n|O) \quad \times$$

$$\times \sum_{l=0}^{\infty} \frac{(-1)^l}{2^l l!} \rho^l =$$

$$= e^{-\frac{1}{2}\rho} \prod_{r=1}^{\infty} \frac{1}{\sqrt{n_r!}} \{(-i)(2\pi)^{5/4} \sqrt{\Delta_r/2\omega_r} \times$$

$$\times J^{\mu}(k_r) e_{\mu}^{(\lambda_r)}\}^{n_r} .$$

The cross-section for the process $\Phi \rightarrow \Psi +$ n soft photons is proportional to the square of the absolute value of the matrix element summed over the polarizations:

$$\sum_{pol} |(\Psi, n\gamma^S|\Phi)|^2 = |(\Psi|S^{\Psi}|\Phi)|^2 e^{-\rho} \prod_{r=1}^{\infty} \frac{\rho_r^{n_r}}{n_r!} ;$$

with

$$\rho_r = \sum_{pol} (2\pi)^{5/2} \sqrt{\Delta_r/2\omega_r} \sqrt{\Delta_s/2\omega_s} \times$$

$$\times J^{\mu}(r) J^{\nu *}(s) \epsilon_{\mu}(r) \epsilon_{\nu}(s) =$$

$$= (2\pi)^{5/2} \Delta_r/2\omega_r |\vec{J}(k_r)|^2 .$$

Obviously we have (compare (2,5))

$$\sum_r \rho_r = \rho ,$$

and since

$$(\rho_1 + \rho_2 + \ldots + \rho_s)^n =$$

$$= \sum_{n_1+n_2+\ldots+n_s=n} \frac{n!}{n_1!\ldots n_s!} \rho_1^{n_1} \rho_2^{n_2} \ldots \rho_s^{n_s} ,$$

the summation over different n_r, taking into account the condition $\Sigma n_r = n$, leads to a Poisson distribution for the factor due to soft photons,

$$\sum_{n_r} \sum_{pol} |(\Psi; n_1, n_2, \ldots |S|\Phi)|^2 =$$

$$= |(\Psi|S^\Psi|\Phi)|^2 e^{-\rho} \sum_{n_r} \prod_{r=1}^{\infty} \frac{1}{n!} \frac{n!}{n_r!} \rho_r^{n_r} =$$

$$= |(\Psi|S^\Psi|\Phi)|^2 e^{-\rho} \frac{\rho^n}{n!} . \qquad (2,8)$$

The number of emitted soft photons is undefined, we still have to sum expression (2,8) over n from O to ∞. Then the divergences cancel in the total factor due to soft photons, leading to ($e^{-\rho} e^{\rho} = 1$)

$$\sum_{\text{all}\,\gamma^s} |(\Psi, \gamma^s |S|\Phi)|^2 = |(\Psi|S^\Psi|\Phi)|^2 . \qquad (2,9)$$

It should be kept in mind that in S^Ψ we have $\omega > \epsilon$, so that (2,9) represents a finite (not infrared-divergent) quantity. The result is that the infinite number of soft photons has no measurable effect on the cross-section. This, however, holds true only in a limited sense, i.e. as long as these soft photons carry a vanishingly small energy with them. How to evaluate the sum in (2,9) for real processes has been shown in chapter I: one introduces a subsidiary condition representing the fact that the undetected quanta carry away a certain finite four-momentum determined by the experimental resolution.

4. Summation over Soft Photons

In analogy to section I.3 we now sum over all soft photons under the above-mentioned subsidiary condition for which we assume that the photons altogether can only carry with them a certain maximum energy:

$$\sum_{r=1}^{\infty} n_r \omega_r \leq \Delta E \ .$$

This represents the most important special case of the general condition. Part of the calculation already has been done in section I.3, here we shall also go into the mathematical details.

The above condition can be incorporated by means of a step function as

$$I = \int_0^{\Delta E} \delta(\sum_r n_r \omega_r - x)\, dx =$$

$$= \int_0^{\Delta E} dx \int_{-\infty}^{+\infty} \frac{dy}{2\pi} e^{-i(\sum_r n_r \omega_r - x) y} =$$

$$= \frac{1}{2\pi} \int_0^{\Delta E} dx \int_{-\infty}^{+\infty} dy\, e^{ixy} \prod_{r=1}^{\infty} e^{-i n_r \omega_r y} \ . \tag{2,10}$$

Thus we get for our transition probability, assuming an "energy uncertainty" ΔE:

$$\sum_n |(\Psi|S|\Phi)|^2 = |(\Psi|S^{\Psi}|\Phi)|^2 \sum_n \{ e^{-\rho} \prod_{r=1}^{\infty} \frac{\rho_r^{n_r}}{n_r!} I \} \ . \tag{2,11}$$

The second factor in (2,11), newly added to (2,9), represents the influence of the energy uncertainty ΔE. It essentially depends on $\vec{J}$ via ρ_r and therefore contains (compare (1,19) and (2,5)) the quantity β and of course the cutoff ϵ. We abbreviate it by

$$b = \frac{1}{2\pi} \int_0^{\Delta E} dx \int_{-\infty}^{+\infty} dy \, e^{ixy} \, e^{G(y)} = b(\Delta E, \epsilon, \beta)$$

where (since $\sum_r \rho_r = \rho$)

$$e^{G(y)} = \sum_{n_r=0}^{\infty} \prod_{r=1}^{\text{limit}} e^{-\rho_r} \frac{\rho_r^{n_r}}{n_r!} e^{-in_r \omega_r y} =$$

$$= \prod_{r=1}^{\text{limit}} e^{-\rho_r} e^{\rho_r \exp(-i\omega_r y)} .$$

Here the summation is not extended to infinitely large momenta k_r, which would contradict the concept of soft photons, but only to a k_{limit} conveniently assumed of the order of ϵ. Therefore $G(y)$ is given by

$$G(y) = \sum_{r=1}^{\text{limit}} \rho_r (e^{-i\omega_r y} - 1) \rightarrow$$

$$2\pi \int_0^{\epsilon} \frac{d^3k}{2\pi} |\vec{J}(\vec{k})|^2 (e^{-i\omega y} - 1) =$$

$$= \beta \int_0^{\epsilon} \frac{d\omega}{\omega} (e^{-i\omega y} - 1) .$$

The additional factor b has the form

$$b = \frac{1}{2\pi} \int_0^{\Delta E} dx \int_{-\infty}^{+\infty} dy \exp\left\{ ixy + \beta \int_0^{\epsilon} \frac{d\omega}{\omega}(e^{-i\omega y} - 1) \right\} . \tag{2,12}$$

Obviously we get $b = 0$ for $\Delta E \equiv 0$, a result already known: a purely elastic scattering cannot occur. In the following we shall evaluate b for $0 < \Delta E \leq \epsilon$. At first we compute the last integral:

$$D = \beta \int_0^{\epsilon} \frac{d\omega}{\omega}(e^{-i\omega y} - 1) =$$

$$= \lim_{\eta \to 0} \beta \left[\int_{\eta y}^{\epsilon y} \frac{\cos z}{z} dz - \right.$$

$$\left. - i \int_{\eta y}^{\epsilon y} \frac{\sin z}{z} dz - \ln\epsilon + \ln\eta \right] .$$

These integrals are known as Cosine and Sine integrals, viz.

$$Ci(\alpha) = - \int_{\alpha}^{\infty} \cos z/z \, dz \quad ;$$

$$Si(\alpha) = \int_0^{\alpha} \sin z/z \, dz \quad ;$$

$$\lim_{\alpha \to 0} Ci(\alpha) = \ln\gamma + \ln\alpha, \quad (\ln\gamma \ldots \text{ Euler's constant}) ;$$

$$\lim_{\alpha \to \infty} Si(\alpha) = \frac{\pi}{2} .$$

Therefore

$$D = \lim_{\eta \to 0} \beta [- Ci(\eta y) + Ci(\epsilon y) - iSi(\epsilon y) - \ln\epsilon + \ln\eta] =$$

$$= \beta [- \ln\gamma - \ln y - \ln\epsilon + Ci(\epsilon y) - iSi(\epsilon y)] .$$

The functions Ci and Si are not very suitable for further integration, we shall therefore eliminate them formally by introducing an auxiliary function Δ:

$$\Delta = \beta \int_{\epsilon}^{\infty} \frac{d\omega}{\omega} e^{-i\omega y} = \beta [- \mathrm{Ci}(\epsilon y) + i \, \mathrm{Si}(\epsilon y) - i \frac{\pi}{2}] ;$$

and adding it to D to get

$$\hat{D} = D + \Delta = - \beta [\ln\gamma + \ln \epsilon y + i \frac{\pi}{2}] .$$

Now we rewrite the integral in (2,12) as

$$b = b_1 + b_2 = \frac{1}{2\pi} \int_0^{\Delta E} dx \int_{-\infty}^{+\infty} dy \; e^{ixy} e^{\hat{D}} +$$

$$+ \frac{1}{2\pi} \int_0^{\Delta E} dx \int_{-\infty}^{+\infty} dy \; e^{ixy} (e^{\hat{D} - \Delta} - e^{\hat{D}}) .$$

The advantage of this method lies in the fact that under certain conditions b_2 vanishes while b_1 can be calculated easily.

As next step we show: $b_2 = O$ if $\frac{\Delta E}{\epsilon} \leq 1$.

$$2\pi e^{\beta(\ln\gamma + i\frac{\pi}{2})} b_2 =$$

$$= \int_0^{\Delta E} dx \int_{-\infty}^{+\infty} dy \; e^{ixy} e^{-\beta \ln(\epsilon y)} (e^{-\beta \int_{\epsilon}^{\infty} \frac{d\omega}{\omega} e^{-i\omega y}} - 1) =$$

$$\underset{\substack{\eta = \epsilon y \\ z = \frac{\omega}{\epsilon}}}{=} \int_0^{\Delta E} dx \int_{-\infty}^{+\infty} \frac{d\eta}{\epsilon} e^{i\frac{x}{\epsilon}\eta} \eta^{-\beta} (e^{-\beta \int_1^{\infty} \frac{dz}{z} e^{-i\eta z}} - 1) =$$

$$\underset{\left\{\substack{\xi=\frac{x}{\epsilon} \\ t=-i\eta z}\right\}}{=} \frac{\Delta E}{\epsilon} \int_0^{\epsilon} d\xi \int_{-\infty}^{+\infty} \frac{d\eta}{\eta^{\beta}} e^{i\eta\xi} \left(e^{-\beta \int_{-i\eta}^{\infty} \frac{e^t}{t} dt} - 1\right) .$$

Again we employ the theory of functions of complex variables to arrive at statements on the function known as exponential integral

$$\mathrm{Ei}(z) = \int_{\infty}^{z} \frac{e^t}{t}\, dt .$$

We find that Ei (z) is analytic as long as the path of integration is such that $\frac{\pi}{2} \leq \lim_{t \to \infty} \arg t \leq \frac{3\pi}{2}$ and Re t is bounded to the right. Since we have $z = -i\eta$ this implies that $\mathrm{Im}\eta$ is to be bounded to the right, thus we choose especially $\mathrm{Im}\,\eta < 0$.

For $\mathrm{Im}\eta < 0$ and $|\eta| \to \infty$ the factor $e^{i\eta\xi}$ goes to infinity because of the factor $[\exp\{\beta \mathrm{Ei}(-i\eta)\} - 1]$, however, the absolute value of the integrand vanishes in the lower half plane for $|\eta| \to \infty$, we can close the path of integration and get $b_2 = 0$. We obtain the conditions for this result by means of a more detailed investigation:

$$\eta = \mathrm{Re}\eta + i\,\mathrm{Im}\eta \to -iK$$

$$\mathrm{Im}\eta = -K$$

In the calculation $\mathrm{Re}\eta$ is unessential and put to zero. For large η the integrand is therefore

$$\lim_{K \to \infty} K^{-\beta} e^{K\xi} \left(e^{\beta \mathrm{Ei}(-K)} - 1\right) = \lim_{K \to \infty} \frac{\exp\{-\frac{\beta}{K} e^{-K}\} - 1}{K^{\beta} e^{-K\xi}} =$$

$$= \lim_{K\to\infty} \frac{\beta\left(\frac{1}{K}+\frac{1}{K^2}\right) e^{-\frac{\beta}{K} e^{-K}} e^{-K}}{(\beta K^{\beta-1} - \xi K^{\beta}) e^{-K\xi}} =$$

$$= \lim_{K\to\infty} \frac{\left(1+\frac{1}{K}\right) e^{-K(1-\xi)}}{K^{\beta+1}\left(\frac{1}{K}-\xi\right)} = 0 \ ,$$

for $\xi \leq 1$. This, however, implies that b_2 vanishes whenever $\xi_{max} = \frac{\Delta E}{\epsilon} \leq 1$ as stated above.

Therefore we only have to evaluate the first term b_1. We again employ the theory of functions of complex variables, but first we make some rearrangements:

$$b_1 = \frac{1}{2\pi} \int_0^{\Delta E} dx \int_{-\infty}^{+\infty} dy \, e^{ixy} \, e^{\hat{D}} =$$

$$= \frac{1}{2\pi} e^{-\beta(\ln\gamma + i\frac{\pi}{2})} \int_0^{\Delta E} \frac{dx}{\epsilon} \int_{-\infty}^{+\infty} \frac{d\eta}{\eta^{\beta}} e^{i\frac{x}{\epsilon}\eta} =$$

$$= \frac{1}{2\pi} e^{-\beta(\ln\gamma + i\frac{\pi}{2})} i^{\beta-1} \int_0^{\frac{\Delta E}{\epsilon}} d\xi \int_{-i\infty}^{+i\infty} \frac{dt}{t^{\beta}} e^{t\xi} =$$

$$= \frac{e^{-\beta(\ln\gamma + i\frac{\pi}{2})}}{2\pi(-i)^{1-\beta}} \int_0^{\frac{\Delta E}{\epsilon}} \frac{d\xi}{\xi^{1-\beta}} \int_{+i\infty}^{-i\infty} \frac{dr}{r^{\beta}} e^{-r} =$$

$$= \left(\frac{\Delta E}{\epsilon}\right)^{\beta} \frac{e^{-\beta(\ln\gamma + i\frac{\pi}{2})}}{2\pi\beta(-i)^{1-\beta}} \int_{+i\infty}^{-i\infty} \frac{dre^{-r}}{r^{\beta}} \ .$$

After the elementary ξ integration also the remaining integration over r can be executed. We conveniently choose our path of integration in the complex r-plane as shown in Fig. 23.

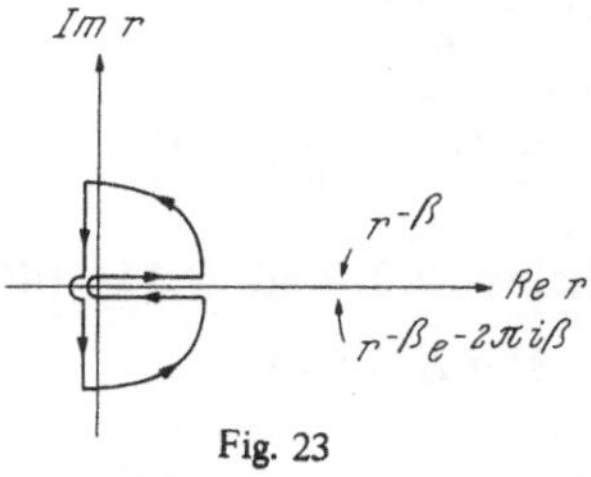

Fig. 23

Because of $r^{-\beta}$ $(0<\beta<1)$ this plane is cut on the real axis from zero to infinity. The integral over the total contour is zero; therefore

$$\int_{+i\infty}^{-i\infty} \frac{dr\, e^{-r}}{r^{\beta}} = -\int_{0}^{\infty} \frac{dr\, e^{-r}}{r^{\beta}} (1-e^{-2\pi i\beta}) =$$

$$= -\Gamma(1-\beta)\,(1-e^{-2\pi i\beta}) = -\frac{2i}{e^{i\pi\beta}}\,\Gamma(1-\beta)\sin\pi\beta =$$

$$= -\frac{(2\pi i)}{(-1)^{\beta}}\,\frac{1}{\Gamma(\beta)} \quad .$$

Taken altogether we have

$$b_1 = \left(\frac{\Delta E}{\epsilon}\right)^{\beta} \frac{\gamma^{-\beta} i^{-\beta} i^{1-\beta}}{2\pi\beta}\,\frac{2\pi(-i)}{(-1)^{\beta}}\,\frac{1}{\Gamma(\beta)} =$$

$$= \left(\frac{\Delta E}{\epsilon}\right)^{\beta} \frac{e^{-\beta\ln\gamma}}{\beta\,\Gamma(\beta)} \quad .$$

Inserting this into (2,12) and taking into account $b_2 = 0$ for $\Delta E \le \epsilon$ we finally get for (2,12)

$$b = b_1 + b_2 = \left(\frac{\Delta E}{\epsilon}\right)^{\beta} \frac{e^{-\beta\ln\gamma}}{\Gamma(1+\beta)} \quad .$$

This also furnishes the proof of (1,20). If we now expand both exponential and gamma-function in a Taylor series we see that the first term, different from one, is of order β^2:

$$\frac{e^{-\beta \ln \gamma}}{\Gamma(1+\beta)} = \frac{1-\beta C+\frac{\beta^2}{2}C^2+\dots}{1-\beta C+\frac{\beta^2}{2}C^2+\frac{\beta^2}{2}\frac{\pi^2}{6}+\dots} \cong 1-\frac{\pi^2}{12}\beta^2 .$$

Therefore the probabilities (2,9) and (2,11) differ essentially in the factor $(\Delta E/\epsilon)^{\beta} \cong (1+\beta \ln \Delta E/\epsilon)$:

$$\sum_n |\Psi, n, (\Delta E)|S|\Phi)|^2 = \left(\frac{\Delta E}{\epsilon}\right)^{\beta} \frac{e^{-\beta \ln \gamma}}{\Gamma(1+\beta)} |(\Psi|S^{\Psi}|\Phi)|^2 . \tag{2,13}$$

Thus we have reached the goal of this chapter: the contributions of soft photons have been incorporated in a single factor of simple form. The apparent dependence of the probability (2,13) on the cut-off ϵ is only artificial; also the matrix element of hard photons has to contain a factor ϵ^{β}. We therefore come back to the discussion following (1,21), now, however, the terms are more clearly defined. To evaluate explicitly the contribution of hard photons for special processes one has to apply the perturbation theory. The results to lowest order in the coupling constant are characterized by the following structure:

$$|(\Psi|S^{\Psi}|\Phi)|^2 = \left(\frac{\epsilon}{E}\right)^{\beta} |M^{(0)}|^2 \{1+\alpha \chi^{(1)}\} ,$$

where $M^{(0)}$ represents the basic process. In case of electron scattering in an external field $\chi^{(1)}$ approximately is of the form:

$$\chi^{(1)} \cong \frac{4}{\pi}\left\{\frac{13}{12}\left(\tfrac{1}{2}\ln \frac{q^2}{m^2}-\tfrac{1}{2}\right)+\frac{17}{72}-\tfrac{1}{2}\sin^2\frac{\Theta}{2} f(\Theta)\right\} .$$

Approximations to higher order lead to very considerable mathematical difficulties; still we may hope that the essential effect of these higher orders is incorporated in the factor $(\frac{\epsilon}{E})^{\beta}$. This factor has been deduced from calculations to lowest order and then modified so that it minimizes the error introduced by the lack of knowledge of the exact contributions from higher orders.

This short repetition of the discussion of chapter I finally yields, starting from (2,13), the often used formula (1,21):

$$d^2\sigma_{exp}(\Delta E) = e^{\beta \ln \Delta E/E}\{\frac{e^{-\beta \ln \gamma}}{\Gamma(1+\beta)}[1+\alpha\chi^{(1)}+\ldots]\}d^2\sigma_o .$$

The first factor essentially represents the influence of the soft photons; the modification (replacement of ϵ by E) serves the purpose that the second factor (in curly brackets) characterizing the hard photons shall differ only a little from unity.

III. Radiative Corrections in the Framework of Quantumelectrodynamics

Calculations in quantumelectrodynamics usually employ the methods of perturbation theory and since the expansion parameter is small the results of lowest order are in general sufficiently exact. In the following we shall therefore consider a simple scattering process with just one additional photon and get acquainted with the occurrence of infrared divergences and the method of how to master their difficulties. These divergences exactly cancel (in any order) if one considers elastic and inelastic events together. Although higher order approximations in QED are connected with seemingly unsurmountable mathematical complications it is, however, possible to sum over the contributions of soft photons in closed form. We then describe in more detail a formalism which can be applied to scattering processes in general. Some special examples will be discussed in chapter IV.

1. Infrared Divergences

As an example we shall consider the simple case of potential scattering. The process is represented in the Feynman-Dyson formalism by the graph

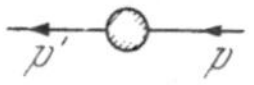

Fig. 24

where the "bubble" indicates the interaction of the electron with the external field (including all possible corrections) and of course we have $E = E'$. The corresponding matrix element has the form

$$M_0(p', p) = \bar{u}(p')\Gamma(p', p)\, u(p) \quad .$$

Here Γ represents the above mentioned bubble; it contains Dirac matrices and therefore is a 4×4 matrix in spinor space but otherwise a scalar.

We now assume the addition of a virtual photon (perturbation theory): then the matrix element is (omitting unessential factors)

$$M_1(p', p) \sim \int d^4k \,\frac{1}{k^2}\, \bar{u}(p')\gamma_\lambda \,\frac{1}{\not{p}' - \not{k} - m} \times$$

$$\times\ \Gamma(p'-k,\ p-k)\ \frac{1}{\not{p} - \not{k} - m}\ \gamma_\mu\ u(p) \ ,$$

as can be easily deduced from the corresponding graph.

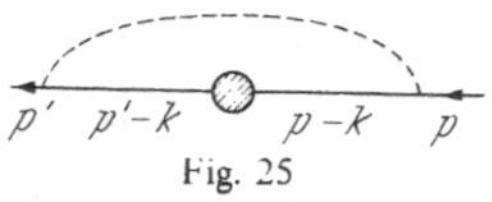

Fig. 25

With the help of the Dirac equation

$$(\not{p} - m)\, u(p) = (p_\mu \gamma^\mu - m)\, u(p) = 0 \ ,$$

we can write M_1 in another form by means of

$$\frac{1}{\not{p} - \not{k} - m}\ \gamma_\mu\, u(p) = \frac{1}{k^2 - 2k \cdot p}\ \{\gamma_\mu m + \not{p}\,\gamma_\mu - \not{k}\,\gamma_\mu\}\, u(p) =$$

$$= \frac{1}{k^2 - 2k\cdot p}\{\gamma_\mu m - \gamma_\mu \not{p} + 2p_\mu - k_\mu - \tfrac{1}{2}[\not{k}, \gamma_\mu]\}u(p) =$$

$$= \frac{2p_\mu - k_\mu - \tfrac{1}{2}[\not{k}, \gamma_\mu]}{k^2 - 2k\cdot p}\, u(p)\ ;$$

and similarly

$$\bar{u}(p')\gamma_\lambda \frac{1}{\not{p}' - \not{k} - m} = \bar{u}(p') \frac{2p'_\mu - k_\mu - \tfrac{1}{2}[\not{k}, \gamma_\mu]}{k^2 - 2k\cdot p'} .$$

This decomposition corresponds to a distinction between effects connected with the spin of the electron (terms with γ) and those which come about even if it has no spin. Therefore we speak of a decomposition into spin and convection terms. The cross-section then contains pure spin, pure convection and mixed spin-convection terms. Also mathematically we thereby have obtained a simplification: taking just a convection term, M_1 is given by

$$M_1 \sim \int \frac{d^4k}{k^2}\, \bar{u}(p')\, \Gamma u(p)\, \frac{(2p-k)\cdot(2p'-k)}{(k^2 - 2k\cdot p)(k^2 - 2k\cdot p')} . \qquad (3,1)$$

As $k \to o$ this expression goes to infinity, whereas in the spin terms this infrared divergence is avoided due to an additional k in the numerator. Since we are just considering the corrections of the basic process M_o only the most important terms will be discussed; therefore in the following we shall be concerned with the infrared divergent term (3,1) and neglect the spin terms, which very much facilitates the calculation.

In (3,1) also an ultraviolet divergence occurs as $k \to \infty$, which, however, can be eliminated if we take into account all other possible graphs to the same order. By virtue of the Ward identity the sum of the contributions of the graphs

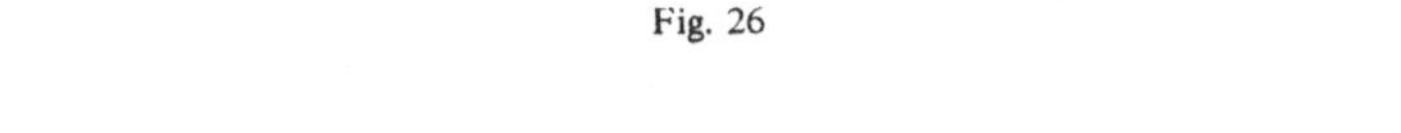

Fig. 26

vanishes for $p = p'$. In our problem we then find that the last factor in (3,1) goes over into

$$\frac{(2p-k)\cdot(2p'-k)}{(k^2-2k\cdot p)\,(k^2-2k\cdot p')} \to -\tfrac{1}{2}\left[\frac{(2p-k)_\mu}{k^2-2k\cdot p} - \frac{(2p'-k)_\mu}{k^2-2k\cdot p'}\right]^2$$

This expression indeed vanishes for $p = p'$ and the integral as a whole exhibits no ultraviolet divergence. Let us now investigate the infrared divergence in (3,1). It can be avoided by introducing a cut-off parameter: one possibility is to integrate down just to a photon energy $\omega_{min} > 0$; in the second method one usually attributes to the photon a small mass λ which corresponds to the substitution

$$\frac{1}{k^2} \to \frac{1}{k^2-\lambda^2} \quad .$$

Whichever cut-off method we may use, however, the parameter will not show up in the final result if we consider both virtual and real photons. The cut-off has to be introduced only to ensure convergence of these contributions which in perturbation theory are treated separately; of course in both the same parameter must be used.

It is interesting to note that this procedure is not just a mathematical artifice but has a physical justification. If a charged particle is deflected its self field has to be deformed according to the change in motion. In order to achieve this deformation it takes a finite time. If we observe the particle after a long time T the greater part of the new self field has been formed already, only the components with very low frequencies $\omega < T^{-1}$ will not have been readjusted. This means that there exists a physically meaningful mechanism for an infrared cut-off. Fortunately the final results do not depend on these details since there the cut-off contributions cancel each other.

Choosing the cut-off λ we get the matrix element in the form

$$M_1 = \alpha B M_0 =$$

$$= \frac{i\alpha}{(2\pi)^3} \int \frac{d^4k}{k^2 - \lambda^2} \left[\frac{(2p' - k)_\mu}{2k \cdot p' - k^2} - \right.$$

$$\left. - \frac{(2p-k)_\mu}{2k \cdot p - k^2} \right]^2 \bar{u}(p') \Gamma(p', p) u(p) \ ; \qquad (3,2)$$

where we took the limit

$$\lim_{k \to 0} \Gamma(p' - k, p - k) = \Gamma(p', p) \ .$$

This procedure physically corresponds to the assumption that the low energy quanta which are of interest here are not sensitive to the details of the scattering process in the bubble.
In this connection one introduces the terminology "external" and "internal" radiative corrections. By neglecting the k-dependence on Γ we obtain external corrections in which we are interested here. They are numerically larger and easier to calculate. The internal corrections are much more difficult to evaluate; they, however, no longer contain infrared divergent contributions. In B we conveniently can neglect k_μ in the numerator compared with p_μ and p'_μ, also in the denominator k^2 with $p \cdot k$ and $p' \cdot k$.

The integral in (3,2) may be evaluated by using Feynman techniques. The details of the calculation are given in chapter IV; here we quote the result for large momentum transfer ($p \cdot p' \gg m^2$)

$$\alpha B = -\frac{\alpha}{2\pi} \left\{ \ln \frac{2p \cdot p'}{m^2} \left[\ln \frac{m^2}{\lambda^2} + \tfrac{1}{2} \ln \frac{2p \cdot p'}{m^2} - \tfrac{1}{2} \right] - \right.$$

$$\left. - \ln \frac{m^2}{\lambda^2} \right\} \ . \qquad (3,3)$$

The cross-section to order α then has to be corrected to

$$\sigma_0 \to \sigma_0 (1 + 2\alpha B) \ . \qquad (3,4)$$

The limit $\lambda \to O$ leads to a meaningless result since it implies $B \to \infty$.

This situation immediately changes if we take into account that due to the experimentally determined observational uncertainties the basic process cannot be distinguished from bremsstrahlung processes with low energy quanta. The cross-section observed therefore consists of (3,4) and the corresponding inelastic contribution.

Also the effects of real photons we include to order α only. From the two graphs

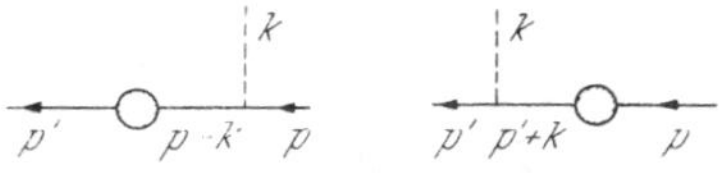

Fig. 27

we construct the inelastic convection-matrix element

$$M_i \sim \left(\frac{(2p-k)_\mu}{k^2 - 2k \cdot p} + \frac{(2p'+k)_\mu}{k^2 + 2k \cdot p'} \right) M_0 \ . \qquad (3,5)$$

An additional term originating from the graph

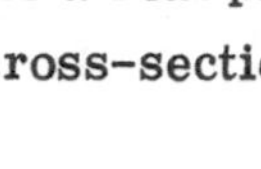

Fig. 28

we neglect since it would be connected with the internal corrections. Expression (3,5) therefore represents the inelastic, infrared divergent amplitude as a whole; for a real photon ($k^2 = O$, $e^\mu k_\mu = O$) we get the bremsstrahlung cross-section

$$\sigma(p, k, p')\, d\Omega_e\, d\Omega_k\, d\omega =$$

$$= - \frac{\alpha}{(2\pi)^2} \omega^2 \left[\frac{p_\mu}{k \cdot p} - \frac{p'_\mu}{k \cdot p'} \right]^2 \sigma_{el}(p, p')\, d\Omega_e\, d\Omega_k \frac{d\omega}{\omega} \ .$$

(3,6)

We have already performed the integration over the photon angle several times; it results in the factor

$$\beta = -\frac{\alpha}{(2\pi)^2}\int d\Omega\omega^2\left[\frac{p_\mu}{k\cdot p}-\frac{p'_\mu}{k\cdot p'}\right]^2 \rightarrow \frac{2\alpha}{\pi}\left[\ln\left(\frac{-q^2}{m^2}\right)-1\right],$$

$$(-q^2) \gg m^2 .$$

The dependence of the cross-section on E' is given by (since $\omega = E - E'$, $d\omega = dE'$)

$$\frac{d\sigma}{dE'} = \frac{\beta}{E-E'}\sigma_{el} .$$

This result again diverges for $k \rightarrow o$ or $E \rightarrow E'$. We therefore insert the same cutoff λ in (3,6) by writing $\omega = \sqrt{\vec{k}^2 + \lambda^2}$. As contribution from real photons which carry away a maximum energy ΔE we then get

$$2\alpha\tilde{B}\,\sigma_o = 2\alpha\frac{(-1)}{2(2\pi)^2}\int_o^{\Delta E}\frac{d^3k}{\sqrt{\vec{k}^2+\lambda^2}}\left(\frac{p_\mu'}{k\cdot p'}-\frac{p_\mu}{k\cdot p}\right)^2\cdot\sigma_o . \quad (3,7)$$

Again we just quote the final result for large q^2.

$$\alpha\tilde{B} = \frac{\alpha}{2\pi}\left\{\ln\frac{2p\cdot p'}{m^2}\left[\ln\frac{m^2}{\lambda^2}+\tfrac{1}{2}\ln\frac{2p\cdot p'}{m^2} - \ln\frac{E^2}{(\Delta E)^2}\right]-\ln\frac{m^2}{\lambda^2}+\ln\frac{E^2}{(\Delta E)^2}\right\} . \quad (3,8)$$

Combining this result with (3,3) and using the quantity β we get for the cross-section to order α

$$\sigma = \sigma_o\left\{1 + 2\alpha(B+\tilde{B})\right\} \cong \sigma_o\left\{1 - \beta\left(\ln\frac{E}{\Delta E}-\frac{1}{4}\right)\right\} , \quad (3,9)$$

which may be compared with (1,36'). As stated before the terms with λ in (3,3) exactly cancel with those in (3,8) and the cross-section depends on the energy uncertainty ΔE only. The result

(3,9), however, can only represent a first order approximation since e.g. for $\Delta E \to o$ the cross-section does not vanish but diverges logarithmically. It is therefore more appropriate to consider (3,9) the beginning of a perturbation series of the form

$$\sigma = \sigma_0 \exp \{ 2\alpha (B + \tilde{B}) \} . \tag{3,9'}$$

As may have been expected from the previously executed classical calculations it is exactly this expression which one arrives at by summing over the soft photon contributions to any order. The details of the quantumelectrodynamical calculation may be found in section III.3.

Here we finally want to write down the generalization of expressions (3,2) and (3,7) for the case that the basic process involves several external (charged) particles. We denote the respective charge of the particles by $e Z_i$, their momenta by p_i, and distinguish by means of the factor $\epsilon_i = \pm 1$ between outgoing $(+1)$ and incoming (-1) particles. The generalization may be accomplished easily through the physical interpretation of the terms

$$\frac{p_\mu}{p \cdot k} - \frac{p'_\mu}{p' \cdot k}$$

as electric currents. We therefore get

$$B = \sum_{\text{pairs}} \frac{-iZ_i \epsilon_i Z_j \epsilon_j}{(2\pi)^3} \int \frac{d^4k}{k^2 - \lambda^2} \left[\frac{(2p_i \epsilon_i - k)_\mu}{k^2 - 2k \cdot p_i \epsilon_i} + \right.$$

$$\left. + \frac{(2p_j \epsilon_j + k)_\mu}{k^2 + 2k \cdot p_j \epsilon_j} \right]^2 , \tag{3,2'}$$

and

$$\tilde{B} = \sum_{\text{pairs}} \frac{Z_i \epsilon_i Z_j \epsilon_j}{2(2\pi)^2} \int_0^{K_m} \frac{d^3k}{\sqrt{\vec{k}^2 + \lambda^2}} \left[\frac{p_{i\mu}}{k \cdot p_i} - \frac{p_{j\mu}}{k \cdot p_j} \right]^2 . \tag{3,7'}$$

Here the summation goes over all external lines. In (3,7') the upper limit K_m depends on the experimental situation; in general K_m may also depend on the photon angle which complicates the integration in (3,7'). In many cases, however, one can find a Lorentz-system where K_m is not longer orientation dependent and the integral can be evaluated easily. The last three formulas represents most useful recipe for the explicit calculation of radiative corrections.

2. The Canceling of Infrared Divergences

We now come back to more general questions: in the following we will show that the divergences cancel not only to order α but to any arbitrary order of perturbation theory.

The basis of the proof is, as explicitly shown in the previous section, the canceling of the divergences (or the cut-off λ) if one considers a soft real and a soft virtual photon in addition to the basic process M_o :

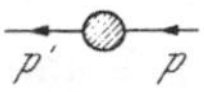

Fig. 29

where the bubble represents the interaction of the charged particle with an external field and, in addition, may correspond to an arbitrary complicated (but not infrared divergent) inner structure.

In the framework of renormalization theory it can be shown that virtual photons which are connected with the interior of the bubble only as

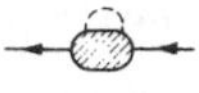

Fig. 30

do not give rise to infrared divergences. This is in accord with the previously discussed view that a photon of extremely long wavelength cannot measure the details of the scattering process. (These "internal" photons have to be taken into account only in the "internal" radiative corrections.) Virtual photons, however, which are also connected with an external line,

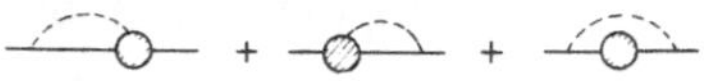

Fig. 31

are the origin of infrared divergences; these are then exactly canceled by the corresponding divergences of the inelastic processes such as

Fig. 32

This result can be generalized in two directions: firstly with respect to the number of soft photons (order α^n) and secondly with respect to the number of incoming or outgoing charged particles. The latter can be understood easily: since the above result holds for each single continuous charged line it must be valid also for more lines participating in a common process:

Fig. 33

To achieve the generalization to arbitrary order in α we proceed conveniently by induction: having proved to order α^n that the divergences cancel we take these graphs as our new basic process and attach an additional photon of which we know that its divergences cancel. Therefore this canceling occurs in any order of the perturbation series. This result could have been expected from the generalization (3,9') of (3,9) (which has still to be established)

$$1 + 2\alpha(B+\tilde{B}) \rightarrow e^{2\alpha(B+\tilde{B})} ,$$

since for the infrared divergent parts we have

$$B_{div} + \tilde{B}_{div} = 0 .$$

Detailed investigation of the process of adding further soft photons, however, reveals some difficulties which we will now briefly discuss: the so-called "overlapping divergences". They occur if simultaneously two photon momenta, connected to the same line, vanish. As the simplest example the following process may be considered:

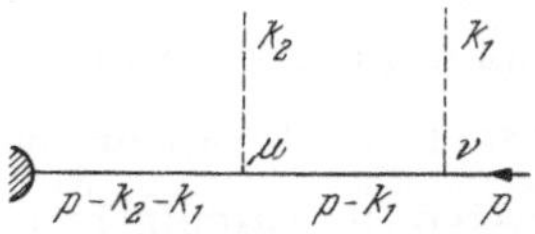

Fig. 34

and we assume that the remaining part of the graph contains no infrared divergences. In the limit $k_1, k_2 \to 0$ this process gives rise to a factor

$$\frac{p_\mu}{p\cdot(k_1+k_2)}\cdot\frac{p_\nu}{p\cdot k_1} \ .$$

There the first term contains both photon momenta, therefore the factorization of the divergences into contributions for each k_i separately cannot be carried out. This difficulty, however, is only artificial since the sum of this expression and the one obtained by exchanging the photon momenta

$$\frac{p_\mu}{p\cdot(k_1+k_2)}\cdot\frac{p_\nu}{p\cdot k_1}+\frac{p_\mu}{p\cdot k_2}\cdot\frac{p_\nu}{p\cdot(k_1+k_2)} =$$

$$=\frac{p_\nu}{p\cdot k_1}\cdot\frac{p_\mu}{p\cdot k_2}$$

obviously is the product of two not overlapping factors.

What we have shown in this special example is valid in general: the proper combination of different terms avoids the occurrence of overlapping infrared divergences and the factorization of the "infrared" terms is therefore always possible.

In the next section we shall confirm these theoretical discussions by explicitly summing over all divergent graphs. The correction factor originating from soft photons will be found to be of the expected exponential form.

3. Summation of the "Infrared" Contributions

We now consider again the possibility of summing over all soft photons. In order to avoid too lengthy expressions we suppress unessential factors - the exact result of this summation we know already [(2,6) and (2,8)]. Also the calculational details will be omitted since they have been discussed earlier.

Our basic process with a predetermined number of external charged lines is represented by the graph

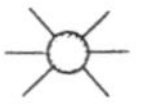

Fig. 35

the corresponding matrix element of which, $\tilde{M}$, contains no soft photons and therefore is not infrared divergent.

Soft photons, however, can be attached to the external lines. Considering for the moment the i-th incoming external line only and assuming that it emits n real soft photons we get as the result of a calculation in analogy to (3,5) and (2,8) the following matrix element:

$$M(n, i) \sim \frac{1}{n!} \sum_{\nu_1 \ldots \nu_n} \tilde{M}_i \frac{1}{\sqrt{2\omega_{\nu_1} \ldots 2\omega_{\nu_n}}} \times$$

$$\times \frac{(\not{p}_i - \not{k}_{\nu_1} - \ldots - \not{k}_{\nu_n} + m_i)}{(p_i - k_{\nu_1} - \ldots - k_{\nu_n})^2 - m_i^2} (e \cdot \gamma) \ldots \times$$

$$\times \ldots \frac{(\not{p}_i - \not{k}_{\nu_1} + m_i)}{(p_i - k_{\nu_1})^2 + m_i^2} (e \cdot \gamma)\, u(p_i) =$$

$$= \frac{1}{n!} \tilde{M} \prod_{\nu=1}^{n} \frac{(e_\nu \cdot p_i)}{\sqrt{2\omega_\nu}(-k_\nu \cdot p_i)} .$$

Here we used $\tilde{M} = \tilde{M}_i \, u(p_i)$ and

$$\sum_{\text{perm}} \{a_1 \cdot (\ )\ldots(a_1 + a_2 + \ldots + a_n)\}^{-1} = \prod_{\nu=1}^{n} \frac{1}{a_\nu};$$

e_ν represents the polarization vector of the ν-th photon. Again we considered the convection terms only in order to achieve the factorization. Generalization to incoming ($\epsilon_i = -1$) and outgoing ($\epsilon_i = +1$) lines leads to

$$M(n, i) \sim \frac{1}{n!} \; \tilde{M} \; \prod_{\nu=1}^{n} \; \frac{(e_\nu \cdot p_i)}{\sqrt{2\omega_\nu}\,(k_\nu \cdot p_i)\,\epsilon_i} \; .$$

More generally we have to find the matrix element $\tilde{M}$ together with l photons exchanged between lines and n photons emitted. The photon ν should be emitted by the line i_ν, the photon λ should be exchanged between lines i_λ and j_λ; this combinatorial problem we already know from (2,7) and (2,8), and we get

$$M(n, l) \; \frac{1}{n!\,l!\,2^l} \; \tilde{M} \left\{ \prod_{\nu=1}^{n} \frac{(p_{i_\nu} \cdot e_\nu)}{\sqrt{2\omega_\nu}\,(k_\nu \cdot p_{i_\nu})\,\epsilon_{i_\nu}} \right\} \times$$

$$\times \prod_{\lambda=1}^{l} \int_{\omega < \epsilon} \frac{d^4 k_\lambda}{k_\lambda^2 - \lambda^2} \; \frac{(p_{i_\lambda} \cdot p_{j_\lambda})}{(k_\lambda \cdot p_{i_\lambda})(-k_\lambda \cdot p_{j_\lambda})\,\epsilon_{i_\lambda}\,\epsilon_{j_\lambda}} .$$

Here ϵ is a measure of the limiting energy of soft photons. We still have to take the sum over the external lines, i.e. over

$$i_1 \ldots i_n,\; i_1 \ldots i_l,\; j_1 \ldots j_l \; .$$

The term resembling a current

$$\sum_{\substack{\text{external} \\ \text{lines}}} \frac{p_i^\mu}{(k \cdot p_i)\,\epsilon_i}$$

we abbreviate by $s^{\mu}(k)$. After the summation over l from o to ∞ we get the matrix element including n additional emitted soft photons (with definite e and $\vec{k}$) and arbitrarily many virtual photons:

$$M(\ldots; e_n, \vec{k}_n) \sim \sum_{l=o}^{\infty} \frac{1}{n!\,l!\,2^l} \left\{ \int_{\omega<\epsilon} \frac{d^4k}{k^2-\lambda^2} s_{\mu}(k)\, s^{\mu}(-k) \right\}^l \times$$

$$\times \prod_{\nu=1}^{n} \frac{(s(k_{\nu}) \cdot e_{\nu})}{\sqrt{2\omega_{\nu}}} \tilde{M} =$$

$$= \frac{1}{n!} \left\{ \prod_{\nu=1}^{n} \frac{(s(k_{\nu}) \cdot e_{\nu})}{\sqrt{2\omega_{\nu}}} \right\} e^{-\frac{1}{2}A(\epsilon)} \cdot \tilde{M} . \quad (3,10)$$

Here we defined (compare the definition of B in (3,2))

$$A(\epsilon) = - \int_{\omega<\epsilon} \frac{d^4k}{k^2-\lambda^2} (s(k) \cdot s(-k)) .$$

In addition we introduce

$$A = - \int \frac{d^4k}{k^2-\lambda^2} (s(k) \cdot s(-k)) ;$$

then the difference $A - A(\epsilon) \sim \int_{\omega>\epsilon}$ certainly contains no infrared divergence.

We now put $n=o$, i.e. we consider virtual soft photons only. We then get for the process with added virtual quanta only the matrix element

$$M \sim e^{-\frac{1}{2}A(\epsilon)} \tilde{M}$$

which by virtue of $A(\epsilon)$ is infrared divergent; this is not the case for the auxiliary quantity

$$\bar{M} \sim e^{\frac{1}{2}(A-A(\epsilon))} \tilde{M} . \quad (3,11)$$

Applying (3,11) we then can write

$$M \sim e^{-\frac{1}{2}A}\,\bar{M}\ . \qquad (3,12)$$

Equation (3,12) represents a very important relation: it demonstrates how to evaluate the matrix element for the basic process plus virtual soft photons. The factor $e^{-\frac{1}{2}A}$ contains the summation over all virtual soft photons whereas the second factor in (3,12) no longer is infrared divergent. It may be calculated explicitly by the following procedure:

Consider M and $\bar{M}$ expanded in a power series in α:

$$M = M_o + \alpha M_1 + \dots\ ,$$

$$\bar{M} = \bar{M}_o + \alpha \bar{M}_1 + \dots\ .$$

If we insert these into (3,12) and observe that the quantity A also is of order α (taking into account all factors A corresponds to αB) we get by comparison

$$\bar{M}_o = M_o\ ,$$

$$\bar{M}_1 = M_1 + \tfrac{1}{2} B\, M_o\ ,$$

$$\bar{M}_2 = \dots\ .$$

This now constitutes a recipe for the calculation of the terms $\bar{M}_i$. In practice one has to terminate always after the second term since the evaluation of the perturbation theoretical expressions M_i involves too many difficulties for $i > 1$. An example for this calculation can be found in chapter IV. There one may see indeed that $\bar{M}_1$ is free from infrared divergences.

The most general case, however, is represented by the matrix element (3,10) with $n \neq o$. If we again introduce the auxiliary matrix element $\bar{M}$, defined as above, we get

$$M(\ldots; e_n \vec{k}_n) \sim \frac{1}{n!} \{ \prod_{\nu=1}^{n} \frac{(s(k_\nu) \cdot e_\nu)}{\sqrt{2\omega_\nu}} \} e^{-\frac{1}{2}A} \bar{M} .$$

From this follows the transition probability for a process with n real soft photons as

$$P(\vec{k}_1, \ldots \vec{k}_n) = \sum_{pol} \sum_{perm} M^+ M = \sum_{pol} n! |M|^2 =$$

$$= \frac{1}{n!} \prod_{\nu=1}^{n} (- \frac{s(k_\nu) \cdot s^*(k_\nu)}{2\omega_\nu} e^{-\mathrm{Re}\, A} |\bar{M}|^2 ,$$

which conforms with (2,8). As demonstrated in chapter I, we may sum over all photons (n = o to ∞), taking into account the subsidiary condition that the photons can carry with them only a certain energy ΔE.

We finally get the formula

$$P(\Delta E) = \sum_{n=0}^{\infty} \frac{1}{n!} \int \ldots \int_{\omega_\nu < \epsilon} d^3k_1 \ldots$$

$$\ldots d^3k_n \prod_{\nu=1}^{n} \frac{-s(k_\nu) \cdot s^*(k_\nu)}{2\omega_\nu} \times$$

$$\times \frac{1}{2\pi} \int_0^{\Delta E} dx \int_{-\infty}^{+\infty} dy \exp \{ i (\sum_{\nu=1}^{n} \omega_\nu - x) y \} \times$$

$$\times e^{-\mathrm{Re}\, A} |\bar{M}|^2 =$$

$$= \frac{1}{2\pi} \int_0^{\Delta E} dx \int_{-\infty}^{+\infty} dy\, e^{ixy} \exp \{ \int_{\omega < \epsilon} \frac{d^3k}{2\omega} (-s(k) \times$$

$$\times\ s^*(k))\, e^{-i\omega y}\}\, e^{-\mathrm{Re}\,A}\, |\bar{M}|^2 .$$

This we can write, using the abbreviation known from (2,12)

$$\frac{1}{2\pi}\int_0^{\Delta E} dx \int_{-\infty}^{+\infty} dy \exp\{ixy + \beta \int_0^{\epsilon} \frac{d\omega}{\omega}(e^{-i\omega y} - 1)\} = b =$$

$$= \left(\frac{\Delta E}{\epsilon}\right)^{\beta} \frac{e^{-\beta \ln \gamma}}{\Gamma(1+\beta)}$$

(where the last equality follows under the condition $\Delta E \leq \epsilon$) in the form

$$P(\Delta E, \epsilon) = b(\Delta E, \epsilon) \exp\{-\mathrm{Re}\,A - \int_{\omega<\epsilon} \frac{d^3k}{2\omega}(s(k)\ \times$$

$$\times\ s^*(k))\}\,|\bar{M}|^2 . \tag{3,13}$$

Here one already sees the structure of the expected final formula (3,9'). Now we put the limiting energy ϵ equal to the energy loss ΔE; then the factor b becomes

$$b(\Delta E = \epsilon) = \frac{e^{-\beta \ln \gamma}}{(1+\beta)} \simeq 1 - \beta^2 \frac{\pi^2}{12} ,$$

which differs little from one, and the important term $\left(\frac{\Delta E}{\epsilon}\right)^{\beta}$ has been "shifted" to the second factor which now reads

$$\exp\{-\mathrm{Re}\,A - \int_{\omega<\Delta E} \frac{d^3k}{2\omega}\, s \cdot s^*\} . \tag{3,14}$$

If we finally insert for the current s and the abbreviation A the actual expressions we get for (3,14)

$$\exp\{2\alpha \sum_{\substack{\text{pairs}\\(ij)}} \left[\frac{Z_i \epsilon_i Z_j \epsilon_j}{(2\pi)^3 i} \int \frac{d^4k}{k^2 - \lambda^2} \frac{4(p_i \cdot p_j)}{(k^2 + 2k\cdot p_i)(k^2 - 2k\cdot p_j)} + \right.$$

$$+ \frac{Z_i \epsilon_i Z_j \epsilon_j}{8\pi^2} \int_0^{\Delta E} \frac{d^3k}{\sqrt{\vec{k}^2 + \lambda^2}} \frac{(p_i \cdot p_j)}{(k \cdot p_i)(k \cdot p_j)} \Bigg] \Bigg\}$$

so that (3,9') with the definitions (3,2') and (3, 7') is verified indeed. In this expression the infrared divergences (or rather the photon mass λ introduced to their avoidance) cancel.

Also the last factor in

$$P(\Delta E) = (1 - \beta^2 \frac{\pi^2}{12}) e^{2\alpha(\mathrm{Re}B + \tilde{B})} |\bar{M}|^2 ,$$

according to the discussion following its definition (3,11) is a finite quantity which can be calculated by means of perturbation theory. In $\bar{M}$ also the contributions of hard photons with $\omega > \epsilon$ have to be taken into account. As mentioned earlier, this formalism has to be applied with some care and will be explained by means of special examples in the following chapter.

IV. Examples

1. Electron-Proton-Scattering

a) The Experimental Situation

The detailed knowledge of the experimental setup is of essential importance for the calculation of radiative corrections since the actual computation has to be fitted to the prevailing situation. Here we choose relatively simple experimental circumstances:

An extremely well-focused monoenergetic (E_1) electron beam impinges on a proton target. We shall distinguish between two cases:

I. Detection of the outgoing electrons,

II. Detection of the recoil protons.

The detecting apparatus is a spectrometer with simultaneous observation of energy and scattering angle. The resolution of the spectrometer is assumed to be as follows: only those particles are counted which have an energy not less than a certain value E_{min} and which are scattered into a fixed angular interval $\Delta\Theta$. This situation is shown in Fig. 36.

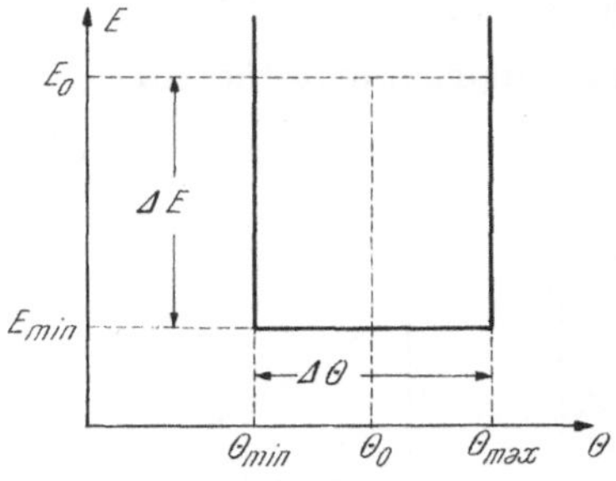

Fig. 36

In this example the particles are said to have an energy E_0 and are scattered about a mean angle Θ_0.

The energy E_0 is uniquely determined by the angle Θ_0, compare I.4 and equation (4,4). From this one would expect just one sharp value for the energy E_o at a fixed Θ_o in case of very good angular measurement (Fig. 37a). Instead the actual intensity distribution has the shape as shown in Fig. 37b since the particles can lose energy due to the emission of soft photons. Then E_{min} will be chosen so that possibly all particles belonging to the energy E_o are counted i.e. at such a value where the intensity I (Fig. 37b) already vanishes.

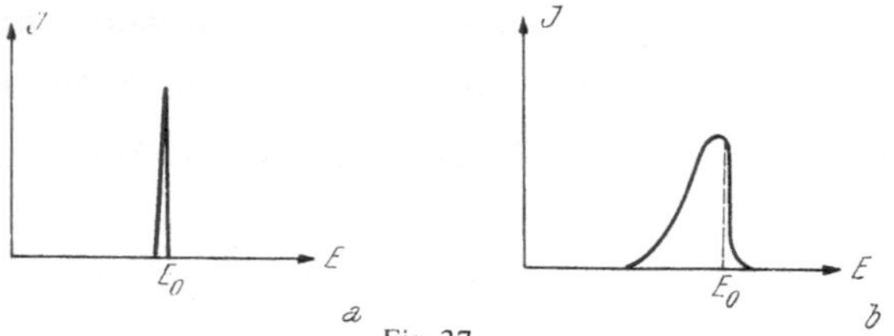

Fig. 37

The fact that also particles with energies larger than E_o occur is attributed to the uncertainties in the primary beam and other sources of error; for reasons of simplicity we assume a perfectly sharp primary beam and therefore a sharp upper limit E_o. For the angular limits Θ_{min} and Θ_{max} we get from equation (4,4) the value E_a and E_b for the energies in case of purely elastic scattering. Therefore equation (4,4) determines the upper boundary of the range ABCD where the particles are observed (Fig. 38), for not too large an interval $\Delta\Theta$ we may approximate the curve $E_a \rightarrow E_o \rightarrow E_b$ by a straight line.

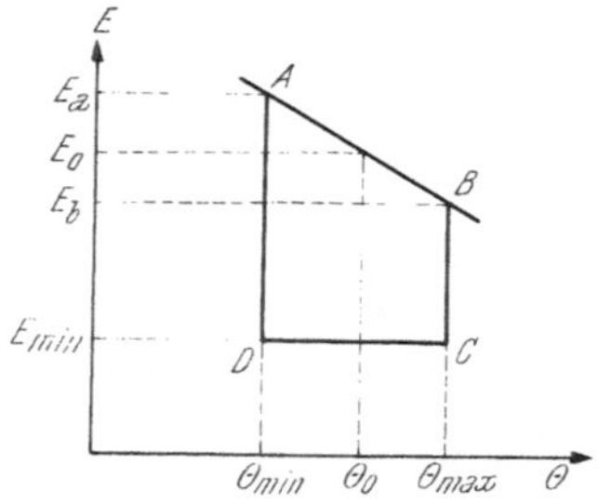

Fig. 38

Here we always take the case of very sharp angular resolution, then the uncertainty in energy $\Delta E = E_o - E_{min}$ is the essential one. According to Fig. 38 we must have $E_b \geq E_{min}$ or

$$\sin^2 \Theta_o \cdot E_o^2 \Delta\Theta \leq 2M \Delta E \ .$$

(Otherwise the uncertainties ΔE and $\Delta\Theta$ would be interchanged in their meaning as essential and secondary uncertainty). Therefore the radiative corrections will depend mainly on ΔE, whereas $\Delta\Theta$ just causes small deviations. In addition to ΔE and $\Delta\Theta$ also the kinematical quantities and their relations are of importance.

b) Kinematics

Our notation can be taken from Fig. 39: a) defines the relevant angles, in b) and c) we define the momenta for the two cases of purely elastic and inelastic scattering.

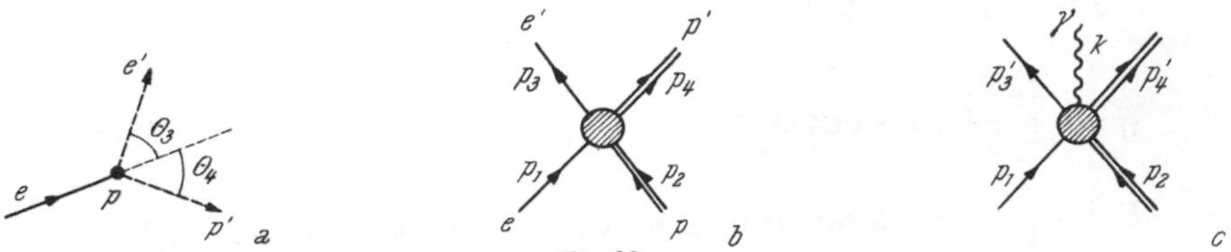

Fig. 39

$$p_1 + p_2 = p_3 + p_4 = p_3' + p_4' + k \qquad (4,1)$$

Another important assumption is that we restrict our investigation to high energy scattering processes, i.e. we may neglect the mass of the electron m in comparison with its energy:

$$p_1^2 = E_1^2 - \vec{p}_1^{\,2} = m_1^2 = m^2, \quad E_1 \simeq P_1 = |\vec{p}_1|$$

$$p_3^2 = \dots = m^2, \qquad E_3 \simeq P_3 \qquad (4,2)$$

The mass of the proton $M = m_2 = m_4$, however, is of the same order of magnitude as the energies under consideration.

The quantities (E_i, P_i, Θ_i) we have just introduced are defined in the lab. system; then p_2 is simply $p_2 = (M; o, o, o)$ and by means of the momentum transfer $q = p_1 - p_3$ we may derive the relation (4,4) between E_1, E_3 and Θ_3, which we already used in I.4:

$$q^2 = (p_1 - p_3)^2 = 2m^2 - 2E_1 E_3 + 2P_1 P_3 \cos\Theta_3 \cong$$

$$\cong - 2E_1 E_3 (1 - \cos\Theta_3) \qquad (4,3)$$

$$q^2 = (p_4 - p_2)^2 = 2M(M - E_4) = -2M(E_1 - E_3) \quad . \qquad (4,3')$$

Equating (4,3) and (4,3') leads to

$$E_3 [M + E_1 (1 - \cos\Theta_3)] = E_1 M$$

or

$$E_3 \eta = E_1$$

$$\eta = 1 + \frac{E_1}{M} (1 - \cos\Theta_3) \quad . \qquad (4,4)$$

For E_1 fixed, a small change $\delta\Theta_3$ of the observation angle results in a change δE_3 of the energy:

$$\delta E_3 = - \frac{E_1}{\eta^2} \frac{E_1}{M} \sin\Theta_3 \, \delta\Theta_3 = \frac{E_3^2}{M} \sin\Theta_3 \, \delta\Theta_3 \quad . \qquad (4,5)$$

From this also the condition follows when ΔE represents the larger uncertainty than $\Delta\Theta$ and furthermore the slope of the "elastic line" AB in Fig. 38.

In addition we want to express E_4 and P_4 in terms of the scattering angle Θ_4:

$$(p_1 - p_4)^2 = (p_2 - p_3)^2 \rightarrow$$

$$m^2 + M^2 - 2E_1 E_4 + 2E_1 P_4 \cos\Theta_4 = m^2 + M^2 - 2ME_3$$

$$E_1 (E_4 - P_4 \cos\Theta_4) = M(E_1 + M - E_4)$$

$$E_4(E_1 + M) = \sqrt{P_4^2 + M^2}(E_1 + M) = E_1 P_4 \cos\Theta_4 + M(E_1 + M) .$$

Solving the last equation for P_4 and E_4 resp. we get:

$$P_4 = \frac{2ME_1 (E_1 + M)\cos\Theta_4}{(E_1 + M)^2 - E_1^2 \cos^2\Theta_4} \tag{4,6}$$

$$E_4 = M \frac{(E_1 + M)^2 + E_1^2 \cos^2\Theta_4}{(E_1 + M)^2 - E_1^2 \cos^2\Theta_4} \tag{4,7}$$

and also

$$\cos\Theta_4 = \frac{P_4}{E_1} \frac{M + E_1}{M + E_4} .$$

These relations we have already used in the discussion following (1,34). By differentiating (4,6) and using (4,7) we obtain the differential relation between δP_4 and $\delta\Theta_4$ as

$$\delta P_4 =$$

$$= - \frac{2ME_1(E_1+M)\sin\Theta_4(\cos\Theta_4/\cos\Theta_4)[(E_1+M)^2 - E_1^2\cos^2\Theta_4 + 2E_1^2\cos^2\Theta_4]}{[(E_1+M)^2 - E_1^2\cos^2\Theta_4]^2}\delta\Theta_4$$

$$\delta P_4 = - \frac{P_4 E_4}{M} \operatorname{tg}\Theta_4 \cdot \delta\Theta_4 . \tag{4,8}$$

Relation (4,8) represents the analog to (4,5) for case II (proton observed) and is used to compare the relative magnitudes of the uncertainties ΔP_4 and $\Delta\Theta_4$ and also to determine the slope of the "elastic line" in the $P_4 - \Theta_4$ diagram.

c) Phase Space of Soft Photons

As we know from the discussion following (1,34) the phase space resulting from an uncertainty ΔE in the energy of the observed electron has the shape of an ellipsoid of revolution. There exists, however, a special coordinate frame where this ellipsoid becomes a sphere; quantities in this system are written as $\tilde{p}_i$.

By taking the square of

$$p_1 + p_2 - p_3' - k = p_4'$$

we get

$$-2k \cdot (p_1 + p_2 - p_3') + 2p_1 p_2 - 2p_1 p_3' - 2p_2 p_3' = 0 \ .$$

In order that the photon angle does not occur we must have

$$\tilde{\vec{p}}_1 + \tilde{\vec{p}}_2 - \tilde{\vec{p}}_3' = \tilde{\vec{p}}_4' + \tilde{\vec{k}} = 0 \ . \tag{4,9}$$

But this special Lorentz frame is exactly the center of mass system of the photon and the unobserved outgoing particle.
In case II the analogous system is characterized by

$$\tilde{\vec{p}}_3' + \tilde{\vec{k}} = 0 \ . \tag{4,9'}$$

Let us now consider case I. The allowed phase space is given by

$$\tilde{\omega}(\tilde{E}_1 + \tilde{E}_2 - \tilde{E}_3') = \tilde{\omega}(\tilde{E}_4' + \tilde{\omega}) = p_2 \cdot (p_1 - p_3') - p_1 \cdot p_3' \ .$$

Since the r.h.s. is the scalar product of 4-vectors it is invariant and can be evaluated in any coordinate system; for this we go back to our original lab. system and get

$$\tilde{\omega}(\tilde{E}_4' + \tilde{\omega}) = ME_1 - ME_3' - E_1 E_3' (1 - \cos\Theta_3) =$$

$$= ME_1 - E_3'[M + E_1(1-\cos\Theta_3) = M\eta E_3 - E_3'M\eta = \eta M\Delta E \quad .$$

Of course this quantity is an invariant, it will be denoted by γ_1:

$$\gamma_1 \equiv k \cdot p_4' = \tilde{\omega}\,(\tilde{E}' + \tilde{\omega}) = \eta M\,\Delta E \quad . \tag{4,10}$$

This relation is exactly valid, however, if $\vec{p}_3$ is parallel to $\vec{p}_3'$ since we assumed $\cos\Theta_3' = \cos\Theta_3$.

In the same way we introduce the invariant

$$\gamma_2 \equiv k \cdot p_3'$$

for case II and get

$$k \cdot p_3' = k \cdot (p_1 + p_2 - p_4') = p_2\,(p_1 - p_4') - p_1\,p_4' + M^2 =$$

$$= ME_1 - ME_4' - E_1\,(E_4' - P_4'\cos\Theta_4) + M^2 =$$

$$= M^2 + ME_1 + M(E_4 - E_4') + E_1\,(E_4 - E_4') - E_4\,(M + E_1) +$$

$$+ E_1\,P_4'\cos\Theta_4 = M^2 + ME_1 + (M + E_1)\,(E_4 - E_4') -$$

$$- E_1\,\Delta P_4\cos\Theta_4 - ME_1 - M^2 =$$

$$= \left(\frac{M + E_1}{E_4}\,P_4 - E_1\cos\Theta_4\right)\Delta P_4 =$$

$$= \frac{(E_1 + M)\,P_4 M}{E_4\,(M + E_4)}\;\Delta P_4$$

or

$$\gamma_2 = \frac{E_1 M}{E_4}\cos\Theta_4\,\Delta P_4 \quad . \tag{4,10'}$$

In order to express $\tilde{\omega}$ in terms of these quantities we employ the relations

$$\tilde{E}'_4 = \sqrt{M^2 + \tilde{k}^2}\ , \quad \tilde{E}'_3 = \sqrt{m^2 + \tilde{k}^2} \quad (\text{where } \tilde{k} = |\tilde{\vec{k}}|)$$

to obtain

$$\tilde{\omega} = \frac{\gamma_1}{\sqrt{M^2 + 2\gamma_1}} \quad \ldots\ldots \text{ case I} \tag{4,11}$$

$$\tilde{\omega} = \frac{\gamma_2}{\sqrt{m^2 + 2\gamma_2}} \quad \ldots\ldots \text{ case II} \quad . \tag{4,11'}$$

This special Lorentz system (with spherical photon-phase-space) enables us to evaluate the photon-phase-space integrals easily.(Unfortunately, problems also occur where this Lorentz system does not exist, as e.g. in e-p scattering when both particles are observed in coincidence. There the phase space is severely deformed and the integration cannot be performed analytically.).

d) Phase-space Integrals

First for the elastic case (without soft photons) under the experimental conditions I:

$$I_1 = \int \frac{d^3P_3}{E_3} \int \frac{d^3P_4}{E_4}\, \delta^4(p_1 + p_2 - p_3 - p_4) =$$

$$= \int \frac{d^3P_3}{E_3} \frac{1}{E_4}\, \delta\,(E_1 + M - E_F)$$

$$E_F = \sqrt{P_3^2 + m^2} + \sqrt{(\vec{p}_1 - \vec{p}_3)^2 + M^2} =$$

$$= E_3 + \sqrt{E_1^2 + E_3^2 - 2E_1 E_3 \cos\Theta_3 + M^2}$$

$$I_1 = d\Omega_3 \int \frac{P_3^2 dP_3}{E_3 E_4} \delta(E_1 + M - E_F) =$$

$$= d\Omega_3 \int \frac{P_3}{E_4} \frac{\partial E_3}{\partial E_F} dE_F \, \delta(E_1 + M - E_F)$$

$$\frac{\partial E_F}{\partial E_3} = 1 + \frac{E_3 - E_1 \cos\Theta_3}{E_4} = \frac{M E_1}{E_4 E_3}, \quad P_3 \approx E_3 \; .$$

$$I_1 = d\Omega_3 \frac{P_3}{E_4} \frac{E_4 P_3}{M E_1} = d\Omega_3 \frac{P_3^2}{M E_1} \; . \tag{4,12}$$

In the same manner we obtain for condition II:

$$I_2 = d\Omega_4 \int \frac{P_4^2 dP_4}{E_3 E_4} \delta(E_1 + M - E_G) =$$

$$= d\Omega_4 \int \frac{P_4^2}{E_3 E_4} \left\{ \frac{\partial P_4^2}{\partial E_G} \right\} dE_G \, \delta(E_1 + M - E_G)$$

$$E_G = E_4 + \sqrt{E_1^2 + P_4^2 - 2 E_1 P_4 \cos\Theta_4}$$

$$\frac{\partial E_G}{\partial P_4} = \frac{P_4}{E_4} + \frac{P_4 - E_1 \cos\Theta_4}{E_3} = \frac{P_4 (E_1 + M) M}{E_3 E_4 (M + E_4)}$$

$$I_2 = d\Omega_4 \frac{P_4}{M} \cdot \frac{M + E_4}{M + E_1} \; . \tag{4,12'}$$

We now turn to the inelastic process:

$$J_1 = d\Omega_3 \int \frac{P_3'^2 dP_3'}{E_3'} \cdot \frac{d^3 P_4'}{E_4'} \cdot \frac{d^3 k}{\omega} \delta(p_1 + p_2 - p_3' - p_4' - k) =$$

$$= d\Omega_3 \int \frac{P_3' \, dP_3'}{E_4'} \cdot \frac{d^3k}{\omega} \, \delta\left(E_1 + E_2 - E_3' - \omega - \sqrt{\vec{p}_4'^{\,2} + M^2}\right)$$

We evaluate this integral in the special Lorentz system where

$$\tilde{\vec{p}}_4' = -\tilde{\vec{k}} \text{ and } \tilde{E}_H = \tilde{E}_3' + \tilde{\omega} + \sqrt{\tilde{k}^2 + M^2}$$

$$\frac{\partial \tilde{E}_H}{\partial \tilde{k}} = \frac{\tilde{k}}{\tilde{\omega}} + \frac{\tilde{k}}{\tilde{E}_4'} = \frac{\tilde{k}}{\tilde{\omega}\tilde{E}_4'} (\tilde{E}_4' + \tilde{\omega}) = \gamma_1 \frac{\tilde{k}}{\tilde{E}_4' \tilde{\omega}^2} \quad .$$

Here we introduced the invariant γ_1 for which we obtained $\gamma_1 =$ $= (P_3 - P_3') M\eta$, and therefore $\partial\gamma_1 / \partial P_3' = - M\eta$.

We then get

$$J_1 = d\Omega_3 \int \frac{P_3'}{\tilde{E}_4'} \frac{\partial P_3'}{\partial \gamma_1} d\gamma_1 \int d\tilde{\Omega}_k \int \frac{\tilde{k}^2}{\tilde{\omega}} \times$$

$$\times \frac{\tilde{E}_4' \tilde{\omega}^2}{\gamma_1 k} \cdot d\tilde{E}_H \; \delta(\tilde{E}_1 + \tilde{E}_2 - \tilde{E}_H)$$

$$J_1 = - d\Omega_3 \int_\Gamma^o \frac{P_3'}{\tilde{E}_4'} \frac{\tilde{E}_4' d\gamma_1}{\gamma_1 \eta M} \int d\tilde{\Omega}_k \, \tilde{\omega} \tilde{k} \simeq$$

$$\simeq d\Omega_3 \frac{P_3^2}{ME_1} \int_o^\Gamma \frac{d\gamma_1}{\gamma_1} \int \tilde{\omega} \tilde{k} \, d\tilde{\Omega}_k \qquad (4,13)$$

The same factor emerges as in the elastic case. One remark about the limits of integration: in rewriting the P_3'-integration as one over $\gamma_1 = M\eta (P_3 - P_3')$ one has to take into account that the maximum value of P_3' is P_3, which corresponds to $\gamma_1 = 0$, whereas the minimum of P_3' is fixed by E_{min}, therefore Γ is connected with the energy resolution ΔE.

The analogous calculation for case II leads to the corresponding result:

$$J_2 = d\Omega_4 \int \frac{P_4'^2 dP_4'}{E_4'} \int \frac{d^3P_3'}{E_3'} \int \frac{d^3k}{\omega} \delta(p_1+p_2-p_3'-p_4'-k) = \ldots$$

$$J_2 \simeq d\Omega_4 \frac{P_4(E_4+M)}{M(E_1+M)} \int_0^{\Gamma} \frac{d\gamma_2}{\gamma_2} \int \tilde{\omega}\tilde{k}\, d\tilde{\Omega}_k \quad . \qquad (4,13')$$

Thus we see that the evaluation of the inelastic contributions to radiative corrections in both cases involves integrals of the type

$$\int \frac{d\gamma}{\gamma} \int \tilde{S}\tilde{k}\tilde{\omega}\, d\tilde{\Omega}_k .$$

Here we introduced a further factor S; so far we have just looked at the kinematical factors, disregarding the matrix element itself. The inelastic contribution as a whole is represented by a sum of terms (compare (3,7')):

$$\tilde{B}_{ij} = + \frac{Z_i Z_j \epsilon_i e_j}{8\pi^2} \int_0^{k_m} \frac{d^3k}{\omega} \frac{2(p_i p_j)}{(kp_i)(kp_j)} \quad .$$

This expression we rewrite in our special Lorentz frame and obtain our final form of the integrals under consideration

$$\tilde{B}_{ij} = - \frac{Z_i Z_j \epsilon_i e_j}{4\pi^2} \int_0^{\Gamma} \frac{d\gamma}{\gamma} \int d\tilde{\Omega}_k\, \tilde{\omega}\tilde{k} \frac{(p_i p_j)}{(kp_i)(kp_j)} \quad . \qquad (4,14)$$

The last factor $\tilde{S}$ is an invariant and thus does not necessarily have to be evaluated in our special Lorentz system.

e) Inelastic Contributions

We now want to evaluate the integral (4,14) displaying all the essential steps in the calculation. If one is just interested in the result this paragraph can be skipped.

For the integral $\int d\tilde{\Omega}\, \tilde{k}\, \tilde{\omega}\, S$ we conveniently use Feynman's method (compare the calculation of β in (1,23))

$$\frac{1}{ab} = 2 \int_{-1}^{+1} \frac{dz}{[a(1-z)+b(1+z)]^2}$$

and introduce for this purpose the auxiliary vector p_z as

$$2p_z = p_i\,(1-z) + p_j\,(1+z)\,.$$

If we denote with Θ the angle between $\vec{\tilde{k}}$ and $\vec{\tilde{p}}_z$ we get (for the sake of simplicity we suppress the notation $\sim$ for the time being)

$$\int d\Omega\, k\, \omega\, S = \tfrac{1}{2} \int_{-1}^{+1} dz\, k\, \omega\, (p_i p_j) \int \frac{d\Omega}{(k \cdot p_z)^2} =$$

$$= \pi \int_{-1}^{+1} dz\, k\, \omega\, (p_i p_j) \int \frac{d\cos\Theta}{(E_z\,\omega - P_z k \cos\Theta)^2} =$$

$$= 2\pi\, k\, \omega\, (p_i p_j) \int_{-1}^{+1} \frac{dz}{E_z^2\,\omega^2 - P_z^2 k^2}\ .$$

The integration over z we shall perform as last step; first we evaluate the integral

$$J = \int_{\gamma_{min}}^{\gamma_{max}} \frac{d\gamma}{\gamma}\, k\, \omega\, \frac{1}{E_z^2\,\omega^2 - P_z^2 k^2}\ .$$

We have to do this carefully; in order to avoid divergent results we introduce a finite photon mass λ. In the limit $\lambda \to o$ some terms vanish, other ones, however, remain which we further have to deal with.

Here we restrict ourselves to the above introduced situation I, i.e. the electron is observed. Then the quantity γ is given by

$$\gamma = (k \cdot p_4') = \tilde{\omega}\,\tilde{E}_4' - \tilde{k}\,\tilde{k}\cos\pi = \sqrt{\tilde{k}^2 + M^2}\,\sqrt{\tilde{k}^2 + \lambda^2} + \tilde{k}^2$$

which means

$$\tilde{k}^2 = \frac{\gamma^2 - \lambda^2 M^2}{\lambda^2 + M^2 + 2\gamma}\,, \quad \tilde{\omega}^2 = \tilde{k}^2 + \lambda^2 = \frac{(\gamma + \lambda^2)^2}{\lambda^2 + M^2 + 2\gamma}\,.$$

With the notation $x = \frac{\gamma}{M\lambda}$ we get

$$J = \int \frac{d\gamma}{\gamma}\; \frac{(\gamma + \lambda^2)\sqrt{\gamma^2 - M^2\lambda^2}}{E_z^2(\gamma + \lambda^2)^2 - P_z^2(\gamma^2 - M^2\lambda^2)} =$$

$$= \int \frac{dx}{x}\; \frac{(x + \frac{\lambda}{M})\sqrt{x^2 - 1}}{E_z^2(x + \frac{\lambda}{M})^2 - P_z^2(x^2 - 1)} =$$

$$= \int \frac{dx\,\sqrt{x^2 - 1}}{E_z^2(x + \frac{\lambda}{M})^2 - P_z^2(x^2 - 1)} +$$

$$+ \frac{\lambda}{M} \int \frac{dx}{x}\; \frac{\sqrt{x^2 - 1}}{\cdots\cdots} = J_1 + \frac{\lambda}{M} J_2\,.$$

The limits of the integration over γ are now $\gamma_{min} = M\lambda$ (corresponding to $\tilde{k} = 0$) and Γ; therefore $x_{min} = 1$, $x_{max} = \frac{\Gamma}{M\lambda}$. In the integrand of J_1 we can neglect λ compared to $x\,M$, with the notation $p_z^2 = E_z^2 - P_z^2$ and the substitution

$$x = \frac{1 + u^2}{1 - u^2}\,, \quad x^2 - 1 = \left(\frac{2u}{1 - u^2}\right)^2,$$

$$dx = \frac{4u}{(1 - u^2)^2}\,du, \quad u_{max} = \frac{\sqrt{1 - (M\lambda/\Gamma)}}{\sqrt{1 + (M\lambda/\Gamma)}}$$

we then get

$$J_1 = 8 \int \frac{u^2 du}{[p_z^2(1+u^2)^2 + P_z^2(1-u^2)^2](1-u^2)} =$$

$$= \frac{8}{E_z^2} \int \frac{u^2 du}{(1-u^2)[u^4 + 2u^2((p_z^2 - P_z^2)/E_z^2) + 1]} =$$

$$= 2/p_z^2 \int du \left\{ \frac{1}{1-u^2} + \right.$$

$$\left. + \frac{u^2 - 1}{u^4 + 2u^2((p_z^2 - P_z^2)/E_z^2) + 1} \right\} .$$

Now we have

$$\frac{2}{p_z^2} \int \frac{du}{1-u^2} = \frac{1}{p_z^2} \int \left(\frac{1}{1+u} + \frac{1}{1-u} \right) du = \frac{1}{p_z^2} \ln \frac{1+u}{1-u} \Big|_0^{u_{max}} =$$

$$[\lambda \to O : u_{max} \to \sqrt{1 - 2(M\lambda/\Gamma)} \propto 1 - (M\lambda/\Gamma)]$$

$$= \frac{1}{p_z^2} \ln \frac{2-(M\lambda/\Gamma)}{(M\lambda/\Gamma)} \simeq \frac{1}{p_z^2} \ln \frac{2\Gamma}{M\lambda}$$

and

$$\frac{2}{p_z^2} \int \frac{du(u^2-1)}{u^4 + 2u^2((p_z^2 - P_z^2)/E_z^2) + 1} =$$

$$= \frac{E_z}{p_z^2 P_z} \int du \; \frac{u - (P_z/E_z)}{u^2 - (2P_z/E_z)u + 1} -$$

$$- \frac{u + (P_z/E_z)}{u^2 + (2P_z/E_z)u + 1}) =$$

$$= \frac{E_z}{2p_z^2 P_z} \ln \frac{u^2 - (2P_z/E_z)u + 1}{u^2 + (2P_z/E_z)u + 1} \Big|_0^{u_{max}} =$$

$$= (\xrightarrow{\lambda \to 0}) = \frac{E_z}{2p_z^2 P_z} \ln \frac{E_z - P_z}{E_z + P_z} .$$

Therefore finally

$$J_1 = \frac{1}{p_z^2} [\ln \frac{2\Gamma}{\lambda M} + \frac{E_z}{2P_z} \ln \frac{E_z - P_z}{E_z + P_z}] .$$

The second integral J_2 we do not have to evaluate explicitly. At most terms of the order $\ln \lambda$ occur and therefore $\frac{\lambda}{M} J_2 \xrightarrow[\lambda \to 0]{} 0$. Taken altogether we get from the inelastic contributions

$$\int \frac{d\gamma}{\gamma} \tilde{k}\, \tilde{\omega} \int d\tilde{\Omega}_k S =$$

$$= 2\pi (p_i p_j) \int_{-1}^{+1} \frac{dz}{p_z^2} [\ln \frac{2\Gamma}{M\lambda} + \frac{E_z}{2P_z} \ln \frac{E_z - P_z}{E_z + P_z}]$$

or, according to (4,14),

$$\tilde{B}_{ij} = - \frac{Z_i Z_j \epsilon_i \epsilon_j}{2\pi} (p_i p_j) \int_{-1}^{+1} \frac{dz}{p_z^2} [\ln \frac{2\Gamma}{\lambda M} +$$

$$+ \frac{E_z}{2P_z} \ln \frac{E_z - P_z}{E_z + P_z}] . \qquad (4,15)$$

Before performing the z-integration we want to discuss the elastic contributions of virtual soft photons.

f) Elastic Contributions

According to (3,2') we have to evaluate integrals of the form

$$B_{ij} = -\frac{i}{\pi^3} Z_i Z_j \int \frac{d^4k}{k^2-\lambda^2} \frac{(p_i p_j)}{(k^2+2\epsilon_i(kp_i))(k^2-2\epsilon_j(kp_j))} . \tag{4,16}$$

By means of the auxiliary vector

$$2p'_z = \epsilon_i p_i (1-z) - \epsilon_j p_j (1+z)$$

we rewrite the k-integral of (4,16) as +

$$J = \int \frac{d^4k}{(k^2-\lambda^2)\, ab} = 2 \int_{-1}^{+1} dz \int \frac{d^4k}{(k^2-\lambda^2)\,[a(1-z)+b(1+z)]^2} =$$

$$= 2 \int_{-1}^{+1} dz \int \frac{d^4k}{(k^2-\lambda^2)\,[2k^2+2k(p_i\,\epsilon_i(1-z)-p_j\,\epsilon_j(1+z))]^2} =$$

$$= \tfrac{1}{2} \int_{-1}^{+1} dz \int \frac{d^4k}{(k^2-\lambda^2)\,[k^2+2(k\cdot p'_z)]^2} .$$

After a translation of the origin of the k-integration by $-xp'_z$ we obtain for it

$$\int \frac{d^4k}{[(k^2-x^2p'^2_z+\lambda^2(x-1)]^3} .$$

By applying the general Beta - function formula

$$\int \frac{(k^2)^{m-2}}{(k^2+a^2)^n} d^4k = \frac{i\pi^2}{(a^2)^{n-m}} B(m, n-m)$$

and especially $m = 2$, $n = 3$ and $B(2,1) = \Gamma(2)\cdot\Gamma(1)/\Gamma(3) = \frac{1}{2}$ we get

+ Once again we use Feynman's formula $\frac{1}{a^2b} = \int_0^1 \frac{2x\,dx}{[x(a-b)+b]^3}$ to obtain

$$J = \int_{-1}^{+1} dz \int_0^1 dz \int \frac{x\,d^4k}{[x(2k.p'_z+\lambda^2)+k^2-\lambda^2]^3}$$

$$J = -i\pi^2 \int_{-1}^{+1} dz \int_0^1 dx \frac{x}{2\,[x^2 p_z'^2 + \lambda^2 (1-x)]} \; .$$

We are interested in the limit $\lambda \to 0$; in this case the integral over x gives

$$\int_0^1 \frac{x\,dx}{x^2 p_z'^2 + \lambda^2 - \lambda^2 x} = \int_0^1 \frac{1}{2p_z'^2} \frac{2p_z'^2 x - \lambda^2}{x^2 p_z'^2 - \lambda^2 x + \lambda^2} dx +$$

$$+ \lambda^2 \int_0^1 \frac{dx}{2p_z'^2 (x^2 \ldots)} \; .$$

The second term vanishes and we are left with

$$\frac{1}{2p_z'^2} \ln (x^2 p_z'^2 - \lambda^2 x + \lambda^2) \Big|_0^1 = \frac{1}{2p_z'^2} \ln \frac{p_z'^2}{\lambda^2} \; .$$

We therefore have obtained

$$J = -\frac{i\pi^2}{4} \int_{-1}^{+1} \frac{dz}{p_z'^2} \ln \frac{p_z'^2}{\lambda^2}$$

$$B_{ij} = -\frac{i}{\pi^3} Z_i Z_j (p_i p_j) \left(-\frac{i\pi^2}{4}\right) \int \frac{dz}{p_z'^2} \ln \frac{p_z'^2}{\lambda^2}$$

$$B_{ij} = -\frac{1}{4\pi} Z_i Z_j (p_i p_j) \int_{-1}^{+1} \frac{dz}{p_z'^2} \ln \frac{p_z'^2}{\lambda^2} \; . \qquad (4,17)$$

Unfortunately, in (4,17) the somewhat complicated vector p_z' occurs, which we have to express in terms of p_z.

a) For $\epsilon_i = -\epsilon_j$ we have

$$p'_z = p_z,\ p'^{\,2}_z = -\epsilon_i\,\epsilon_j\,p_z^2 \quad .$$

b) For $\epsilon_i = \epsilon_j$ and $i \neq j$ we put $z = x^{-1}$ and see that

$$p'^{\,2}_z = \frac{p_x^2}{x^2} \quad .$$

Therefore

$$\int_{-1}^{+1} \frac{dz}{p'^{\,2}_z} \ln \frac{p'^{\,2}_z}{\lambda^2} = - \int_{-1}^{+1} \frac{dx}{p_x^2} \ln \frac{p_x^2}{x^2 \lambda^2} =$$

$$= - \epsilon_i\,\epsilon_j \int_{-1}^{+1} \frac{dz}{p_z^2} + \epsilon_i\,\epsilon_j \int_{-1}^{+1} \frac{dz}{p_z^2} \ln z^2 \quad .$$

The second term here leads to Eulerian functions.
Before going into detail we shall discuss the last case

c) $i = j$, $(\epsilon_i = \epsilon_j) : p'_z = - \epsilon_i\, z p_i = - \epsilon_i\, z p_z$,

$$p'^{\,2}_z = z^2\, p_z^2 = z^2\, p_i^2$$

$$\int_{-1}^{+1} \frac{dz}{p'^{\,2}_z} \ln \frac{p'^{\,2}_z}{\lambda^2} = - \epsilon_i\,\epsilon_j \int_{-1}^{+1} \frac{dz}{p_i^2} \ln \frac{p_i^2}{\lambda^2} +$$

$$+ \epsilon_i\,\epsilon_j \int_{-1}^{+1} \frac{dz}{p_i^2} \ln z^2 \quad .$$

We now combine the results (4,15) and (4,17)

$$\mathrm{Re}\, B_{ij} + \widetilde{B}_{ij} = - \frac{\epsilon_i\,\epsilon_j}{2\pi} (p_i\, p_j)\, Z_i\, Z_j \quad \cdot$$

$$\cdot \int_{-1}^{+1} \frac{dz}{p_z^2} \left[\ln \frac{2\Gamma}{M} - \ln \lambda + \frac{E_z}{2 P_z} \ln \frac{E_z - P_z}{E_z + P_z} - \right.$$

$$- \tfrac{1}{2} \ln \frac{p_z^2}{\lambda^2} + \tfrac{1}{2} \ln z^2 \delta(\epsilon_{i}, \epsilon_j)] \qquad (4,18)$$

where $\delta(\epsilon_i, \epsilon_j) = 1$ if $\epsilon_i = \epsilon_j$ and zero otherwise. As has been noted several times the arbitrarily chosen photon mass λ disappears from the result.

g) Evaluation of the z-integration

Again we first discuss the case $\epsilon_i = - \epsilon_j$. The integrand is of the form

$$\frac{1}{2p_z^2} [\ln \frac{4\Gamma^2}{M^2 p_z^2} + \frac{E_z}{P_z} \ln \frac{E_z - P_z}{E_z + P_z}] = (\text{with } W = \frac{\Gamma}{M}) = \qquad (4,19)$$

$$= \frac{1}{2p_z^2} [\ln \frac{W^2}{E_z^2} + \ln \frac{4E_z^2}{(E_z + P_z)^2} + \frac{E_z - P_z}{2P_z} \ln \frac{E_z - P_z}{E_z + P_z}]$$

By means of graphical integration one can show that the first two terms dominate this expression. We therefore only have to evaluate the integral

$$\int_{-1}^{+1} \frac{dz}{p_z^2} \ln \frac{E_z}{W} =$$

$$= 4 \int_{-1}^{+1} \frac{dz}{z^2 (p_i - p_j)^2 - 2z (p_i^2 - p_j^2) + (p_i + p_j)^2} \times$$

$$\times \ln \frac{E_i + E_j - z(E_i - E_j)}{2W} =$$

$$\doteq \frac{4}{(z_1 - z_2)(p_i - p_j)^2} \int_{-1}^{+1} dz \left(\frac{1}{z - z_1} - \frac{1}{z - z_2} \right) \ln (b + az)$$

where

$$a = -\frac{E_i - E_j}{2W}, \quad b = \frac{E_i + E_j}{2W}, \quad W = \frac{\Gamma}{M}$$

$$z_{1,2} = \frac{p_i^2 - p_j^2 \pm 2\sqrt{(p_i p_j)^2 - p_i^2 p_j^2}}{(p_i - p_j)^2},$$

$$z_1 - z_2 = \frac{4\sqrt{(p_i p_j)^2 - p_i^2 p_j^2}}{(p_i - p_j)^2} .$$

For this we use the formula

$$\int dz \frac{\ln (az+b)}{cz+d} = \frac{1}{c} \cdot \ln \left| b - \frac{ad}{c} \right| \cdot \ln \left| a \frac{cz+d}{ad-bc} \right| - $$

$$- \frac{1}{c} \; \Phi \left\{ a \frac{cz+d}{ad-bc} \right\}$$

where $\Phi (x)$ denotes the Eulerian integral

$$\Phi (x) = \int_x^o \frac{\ln |1-y|}{y} dy .$$

Therefore we get

$$\int_{-1}^{+1} \frac{dz}{p_z^2} \ln \frac{E_z}{W} = \frac{1}{\sqrt{\dots}} \left\{ \left[\ln |b+az_1| \ln \left| a \frac{z - z_1}{az_1+b} \right| - \right.\right.$$

$$\left.\left. - \Phi \left(a \frac{z_1 - z}{b+az_1} \right) \right] - [z_1 \to z_2] \right\} \Big|_{-1}^{+1} =$$

$$= \frac{1}{\sqrt{\dots}} \left\{ \left[\ln |b+az_1| \ln \left| \frac{1-z_1}{1+z_1} \right| - \right.\right.$$

$$\left.\left. - \Phi \left(a \frac{z_1 - 1}{az_1+b} \right) + \Phi \left(a \frac{z_1+1}{az_1+b} \right) \right] - [z_1 \to z_2] \right\} .$$

Taken altogether we have

$$\mathrm{Re}\, B_{ij} + \tilde{B}_{ij} = \frac{\epsilon_i \epsilon_j Z_i Z_j (p_i p_j)}{2\pi \sqrt{(p_i p_j)^2 - p_i^2 p_j^2}} \{ [\ln | \frac{E_i + E_j}{2W} - $$

$$- \frac{E_i - E_j}{2W} z_1 | \ln | \frac{1 - z_1}{1 + z_1} | -$$

$$- \Phi (\frac{E_i - E_j - z_1 (E_i - E_j)}{E_i + E_j - z_1 (E_i - E_j)}) + \Phi (\frac{-(E_i - E_j) - z_1 (E_i - E_j)}{E_i + E_j - z_1 (E_i - E_j)})] \cdot$$

$$- [z_1 \to z_2] \} . \tag{4,20}$$

This expression is valid also in case $\epsilon_i = \epsilon_j$ with the exception that an additional contribution arises from $\int dz p_z^{-2} \ln z$:

$$\int_{-1}^{+1} \frac{dz}{p_z^2} \ln z = \frac{1}{\sqrt{\ldots}} \int_{-1}^{+1} dz \left(\frac{1}{z - z_1} - \frac{1}{z - z_2} \right) \ln z =$$

$$= \frac{1}{\sqrt{\ldots}} \{ [\ln | z_1 | \cdot \ln | \frac{z - z_1}{z_1} | -$$

$$- \Phi (\frac{z - z_1}{- z_1}) - [z_1 \to z_2] \} \Big|_{-1}^{+1}$$

$$\int_{-1}^{+1} \frac{dz}{p_z^2} \ln z = \frac{1}{\sqrt{(p_i p_j)^2 - p_i^2 p_j^2}} \{ [\ln | z_1 | \ln | \frac{1 - z_1}{1 + z_1} | -$$

$$- \Phi (\frac{z_1 - 1}{z_1}) + \Phi (\frac{z_1 + 1}{z_1})] - [z_1 \to z_2] \} .$$

This expression in brackets has to be subtracted from the one in (4,20) according to (4,18). It leads, however, to not too large contributions - no term of the form $\ln W = \ln \frac{\Gamma}{M}$ appears and therefore no term as $\ln \frac{\Delta E}{E}$, which is the essential one, as has been discussed previously - and is often neglected.

Finally we discuss the contributions for $i = j$. They can be computed easily in a direct way:

$$\mathrm{Re}\, B_{ii} + \tilde{B}_{ii} = -\frac{Z_i^2 p_i^2}{4\pi p_i^2} \int_{-1}^{+1} dz \left[\ln \frac{4W^2}{p_i^2} + \right.$$

$$\left. + \frac{E_i}{P_i} \ln \frac{E_i - P_i}{E_i + P_i} \right] \quad .$$

In evaluating these formulas one has to remember that all quantities E_i, P_i are to be taken in the special Lorentz system.

h) Computation of Radiative Corrections

In chapter III, equation (3,9), the radiative corrections - expressed in terms of the quantities B, $\tilde{B}$ as evaluated above - were connected with the cross-section by

$$\sigma_{exp} = (1 + \delta)\, \sigma_o \; , \qquad \delta = 2\alpha\, (\mathrm{Re} B + \tilde{B}) \quad . \tag{4,21}$$

Therefore we are left with the explicit computation of the expressions in (4,20) and their summation. In the case under consideration (e-p scattering with ΔE as the dominating uncertainty of observation) some simplifications are possible.

Let us first consider the quantity

$$z_1 = (p_i - p_j)^{-2} \left(p_i^2 - p_j^2 + 2 \sqrt{(p_i\, p_j)^2 - p_i^2 p_j^2}\right)$$

appearing in the first terms of (4,20) and remember our notation

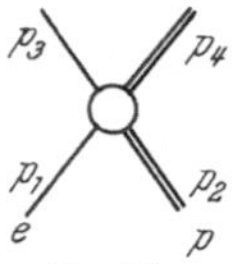

Fig. 40

We discuss some typical examples:

$\boxed{i = 1, \ j = 2}$

$$z_1 = \frac{m^2 - M^2 + 2\sqrt{(E_1 M)^2 - m^2 M^2}}{m^2 + M^2 - 2E_1 M} \simeq \frac{-M + 2E_1}{M - 2E_1}$$

$$z_1 \approx -1 \ .$$

We generally neglect m^2 as compared to M^2 or E_1^2. Obviously this approximation is insufficient in case of

$$\ln \frac{1 - z_1}{1 + z_1} .$$

For $1 + z_1$ we have to go further:

$$1 + z_1 = \frac{2m^2 - 2E_1 M + 2M(E_1^2 - m^2)^{\frac{1}{2}}}{m^2 + M^2 - 2E_1 M} \simeq$$

$$\simeq \frac{2m^2 - M(m^2 / E_1)}{m^2 + M(M - E_1)}$$

$$1 + z_1 \approx \frac{m^2}{ME_1} \frac{2E_1 - M}{M - E_1} .$$

Then we get

$$\ln\left|\frac{\tilde{E}_i + \tilde{E}_j}{2W} - \frac{\tilde{E}_i - \tilde{E}_j}{2W} z_1 \right| \ln\left| \frac{1 - z_1}{1 + z_1}\right| \approx$$

$$\approx \ln\frac{\tilde{E}_1}{W} \cdot \ln\left|\frac{2ME_1}{m^2} \frac{M - 2E_1}{M - E_1}\right| .$$

Here one experiences a difficulty not to be overlooked in numerical computations: for $M = 2E_1$ or $M = E_1$ the term just calculated diverges. It may, however, be shown that by appropriate summation of similar expressions or by more exact calculations these divergences disappear.

$\boxed{i = 1,\ \ j = 3}$

$$(p_1 \cdot p_3) = m^2 - \frac{q^2}{2}, \quad z_1 \simeq 1$$

$$\frac{1-z_1}{1+z_1} \sim \frac{q^2 - q^2(1-4m^2/q^2)^{\frac{1}{2}}}{2q^2} \approx \frac{m^2}{q^2} ;$$

$$\ln\left|\frac{\tilde{E}_i + \tilde{E}_j}{2W} - \frac{\tilde{E}_i - \tilde{E}_j}{2W} z_1\right| \ln\left|\frac{1-z_1}{1+z_1}\right| \approx \ln \frac{\tilde{E}_3}{W} \ln\left|\frac{m^2}{q^2}\right| .$$

Since $W = \frac{\Gamma}{M} = \eta\,\Delta E$ we see that here one of the main contributions appears; considering also the analogous term with z_2 we get

$$z_2 \simeq -1 , \quad (1-z_2)/(1+z_2) = q^2/m^2$$

$$\ln\left\{\frac{1}{2W}\left|\tilde{E}_i + \tilde{E}_j - z_2(\tilde{E}_i - \tilde{E}_j)\right|\right\} \ln\left|\frac{1-z_2}{1+z_2}\right| \simeq$$

$$\simeq \ln \frac{\tilde{E}_1}{W} \ln\left|\frac{q^2}{m^2}\right| .$$

These two terms alone, which have to be subtracted according to (4,20), already give rise to the most essential contribution to δ

$$\delta = 2\alpha\left(-\frac{1}{2\pi}\right)\left\{ \ln\left|\frac{q^2}{m^2}\right| \left| \ln \frac{\eta\,\Delta E}{E_3} + \ln \frac{\eta\,\Delta E}{E_1} \right| \right\} =$$

$$= -\frac{\alpha}{\pi} \ln\left|\frac{q^2}{m^2}\right| \left\{ \ln \frac{\Delta E}{E_3} + \ln \frac{E_1}{E_3} \frac{\Delta E}{E_3} \right\} . \qquad (4,22)$$

This exactly conforms with the main terms in the approximate result of (1,36').

By explicit calculation of the $\tilde{E}_i$ it can be shown that we have the lab. quantities E_1, E_3 occurring in (4,22). One evaluates e.g. the invariant $(p_1 \cdot p_4)$ in both systems:

$$(p_1 \cdot p_4) = \tilde{E}_1 M = p_1^2 + (p_1 p_2) - (p_1 p_3) = m^2 + E_1 M - (p_1 p_3) ;$$

$$\tilde{E}_1 = E_1 - \frac{p_1 p_3}{M}$$

$$\tilde{E}_1 = E_3 .$$

In a similar way we get

$$\tilde{E}_2 = E_4, \quad \tilde{E}_3 = E_1, \quad \tilde{E}_4 = M . \tag{4,23}$$

Therefore equation (4,22) is valid as it stands. We now can obtain more exact results if we add terms such as the one calculated above with $i = 1$, $j = 2$.

We just want to write down one further, also typical, expression of this kind,

$\boxed{i = 2, \; j = 4}$

$$(p_2 \cdot p_4) = ME_4$$

$$z_{1,2} = \pm \frac{2M\sqrt{E_4^2 - M^2}}{2M^2 - 2ME_4} = \mp \frac{P_4}{E_1 - E_3} = \mp \frac{\sqrt{E_1 - E_3 + 2M}}{\sqrt{E_1 - E_3}} .$$

This expression is bounded away from ± 1 and therefore the term $\ln \left| \frac{1 - z_1}{1 + z_1} \right|$ becomes very small. By means of these examples one can see that the main contribution to radiative corrections arises from the light particles (combination 1,3), the next ones being the combinations 1,2 and 1,4 (a light and heavy particle), whereas the heavy particles alone (2,4) contribute least. This result is intuitively plausible from the Fermi - Weizsäcker - Williams approach. Also in (4,20) certain Eulerian integrals occur, the order of magnitude of which also has to be estimated. The explicit calculation of these functions is done by means of their series expansion. We just quote the result in graphical form (Fig. 41).

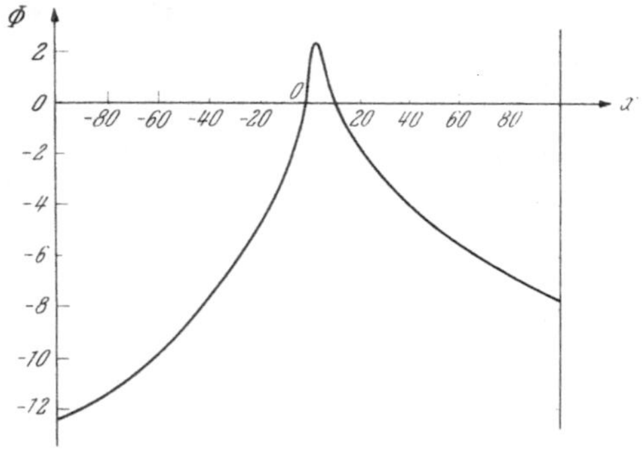

Fig. 41. Euler function Φ (x), -100 ⩽ x ⩽ +100

From the definition (4,19) we see that $\Phi(0) = 0$ and the inequality $|\Phi(x)| < 3$ holds for $|x| < 5$.

Let us consider a typical term of equation (4,20):

$i = 1, \quad j = 3$

$z_1 \approx 1$

$$\Phi = \Phi\left(\frac{-E_i + E_j - (E_i - E_j)\, z_1}{E_i + E_j - z_1(E_i - E_j)}\right).$$

Then

$$\Phi \approx \Phi\left(-\frac{2(\tilde{E}_1 - \tilde{E}_3)}{2\tilde{E}_3}\right) = \Phi\left(1 - \frac{E_3}{E_1}\right).$$

Since $\eta = E_1 / E_3$ is of the order 2 we get $\Phi(\frac{1}{2}) \simeq 0.5$. Multiplied by the factor $\frac{\alpha}{\pi} = \frac{1}{430}$ this term contributes ~0.1 %. Similarly the other Euler functions contribute unessentially to the radiative corrections but should be included in more exact calculations.

The terms we already have neglected in (4,20) are of even less importance generally:

$$\int_{-1}^{+1} \frac{dz}{p_z^2} \ln z \approx \{ \ln |z_1| \ln |\frac{1-z_1}{1+z_1}| - \Phi(\frac{z_1-1}{z_1}) + \Phi(\frac{z_1+1}{z_1}) + \dots \} .$$

In case $i = 1$, $j = 2$ we had $z_1 \cong -1$ and the logarithmic term and the third term actually vanish. Only the second one contributes, again being of order unity.

Finally a remark concerning the actual calculation of radiative corrections. (Until now we have only made estimates of the orders of magnitude which essentially confirmed our previous calculations and may therefore be used as a control of the new formulas.) It would be very tedious to calculate the terms one by one and then to add them, especially if one is interested in radiative corrections for a number of angles and values of energy. Therefore formula (4, 20) will be suitably programmed for an electronic computer, which in the shortest time performs the trivial but tedious steps in the calculation and prints out the final results. They are of the often quoted order of magnitude 10 - 20 %.

i) Matrix Elements

In this paragraph we will put formula (3,11), which so far has been written down only formally, on a practical basis. Therefore we explicitly write down the matrix elements and show how to extract the relevant terms without infrared divergences.

The basic matrix element can be derived from the Feynman graph of Fig. 42 and has the form

$$M_o = \frac{-i\alpha Z_1 Z_2}{\pi} \frac{m M}{\sqrt{E_1 E_2 E_3 E_4}} \times$$

$$\times \frac{1}{q^2} \bar{u}(p_3)\gamma_\mu u(p_1) \bar{u}(p_4)\Gamma^\mu u(p_2) \ .$$

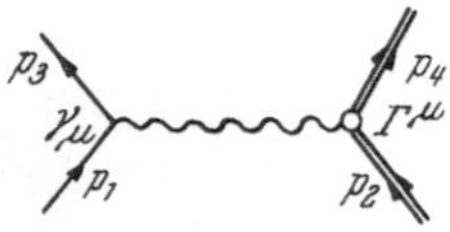

Fig. 42

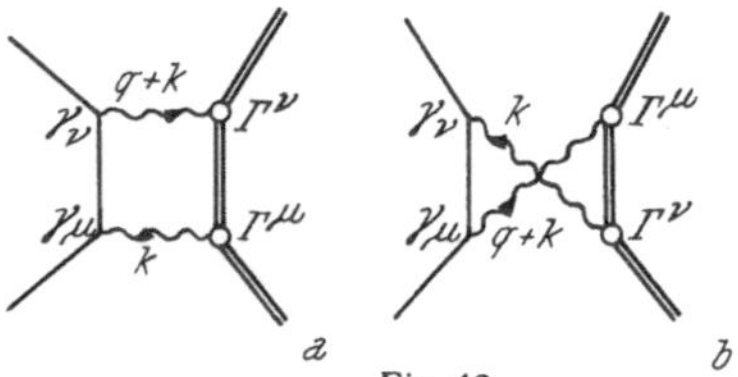

Fig. 43

Here

$$\Gamma_\mu = F_1(q^2)\gamma_\mu + \frac{\varkappa}{2M} F_2(q^2) \not{q}\, \gamma_\mu$$

characterizes the dynamical structure of the proton. Thus it is practically impossible to exactly calculate the two photon exchanges since " off shell form factors " would be involved. For a suitable approximation one considers the infrared divergent contributions of the matrix element M_1^a (Fig. 43a) only:

$$M_1^a = \frac{e^4}{(2\pi)^6} \frac{m M Z_1^2 Z_2^2}{\sqrt{E_1 E_2 E_3 E_4}} \int \frac{d^4k}{(k^2-\lambda^2)((k+q)^2-\lambda^2)} \times$$

$$\times \bar{u}(p_3)\gamma_\nu \{\text{Propagator}\} \gamma_\mu u(p_1) \bar{u}(p_4)\Gamma^\nu \{\text{Propagator}\} \times$$

$$\times \Gamma^\mu u(p_2) \ .$$

Here two types of infrared divergences show up, namely if k or $q + k$ go to zero. Putting $\Gamma = \gamma$ these contributions are

$$M_1^a = \frac{i\alpha Z_1 Z_2}{\pi^3} M_0 \left\{ \int \frac{(p_1 p_2)\, d^4k}{(k^2-\lambda^2)(k^2+2p_1k)(k^2-2p_2k)} + \right.$$

$$\left. + \int \frac{(p_3 p_4)\, d^4k}{(k^2-\lambda^2)(k^2+2p_3k)(k^2-2p_4k)} \right\}.$$

In a similar way we treat diagram Fig. 43 b to get

$$M_1^b = \frac{i\alpha Z_1 Z_2}{\pi^3} M_0 \left\{ \int \frac{(p_2 p_3)\, d^4k}{(k^2-\lambda^2)(k^2-2p_2k)(k^2-2p_3k)} + \right.$$

$$\left. + \int \frac{(p_1 p_4)\, d^4k}{(k^2-\lambda^2)(k^2-2p_1k)(k^2-2p_4k)} \right\} .$$

In the evaluation of the contribution from vacuum polarization diagrams (Fig. 44) can be treated more exactly since we know the vacuum polarization tensor $\Pi_{\mu\nu}$ much better. It has the form (for $|q^2| \gg m^2$)

$$\Pi_{\mu\nu}(q^2) = g_{\mu\nu} q^2 \left\{ - C - \frac{\alpha}{3\pi}\left[\frac{5}{3} - \ln \frac{|q^2|}{m^2} \right] \right\} .$$

The divergent constant C will be incorporated in the charge renormalization and the matrix element then reads

$$M_1^{\text{Vac. Pol.}} = M_0 \frac{\alpha}{\pi}\left(-\frac{5}{9} + \frac{1}{3} \ln \frac{|q^2|}{m^2} \right) .$$

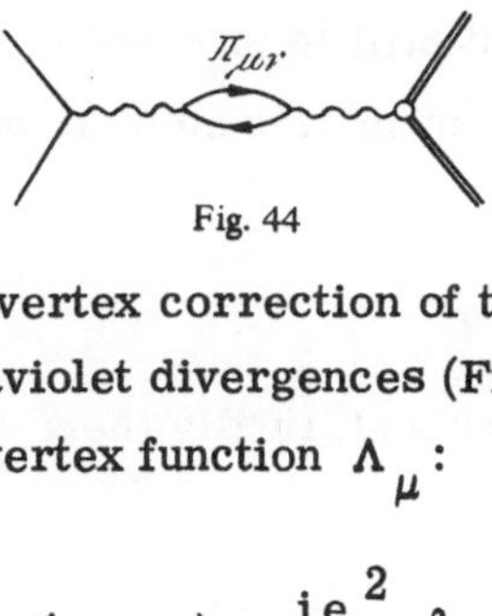

Fig. 44

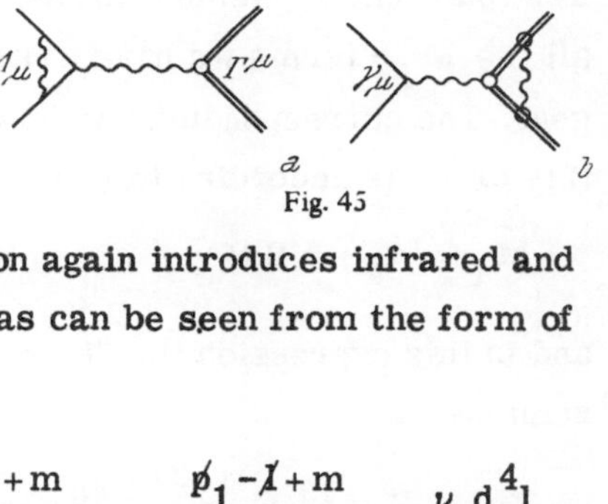

Fig. 45

The vertex correction of the electron again introduces infrared and ultraviolet divergences (Fig. 45a), as can be seen from the form of the vertex function Λ_μ:

$$\Lambda_\mu(p_3, p_1) = \frac{ie^2}{(2\pi)^4} \int \gamma_\nu \frac{\not{p}_3 - \not{l} + m}{(p_3-l)^2 - m^2} \gamma_\mu \frac{\not{p}_1 - \not{l} + m}{(p_1-l)^2 - m^2} \gamma^\nu \frac{d^4 l}{l^2 - \lambda^2} .$$

The ultraviolet divergence can be put incorporated into the renormalization constants, then there remain an infrared divergent and a finite contribution (the latter is written down for $|q^2| \gg m^2$):

$$M_1^{\nu(e)} = M_0 \frac{\alpha}{2\pi}\{-\frac{1}{i\pi^2}\int\frac{d^4 l}{l^2-\lambda^2}\left[\frac{2(p_1p_3)}{(l^2-2p_3l)(l^2-2p_1l)}+\right.$$

$$\left.+\frac{2p_1^2}{(l^2-2p_1l)(l^2+2p_1l)}\right]+\frac{3}{2}\ln\frac{|q^2|}{m^2}-4\} .$$

For the vertex correction to the proton line (Fig. 45b) we would again have to take into account dynamical effects. As an approximation we just consider the infrared divergent contributions which are analogous to the ones of the e-vertex:

$$M_1^{\nu(p)} = -M_0\frac{\alpha Z_2^2}{\pi}\frac{1}{i\pi^2}\int\frac{d^4 l}{l^2-\lambda^2}\left[\frac{(p_2p_4)}{(l^2-2p_4l)(l^2-2p_2l)}+\right.$$

$$\left.+\frac{p_2^2}{(l^2-2p_2l)(l^2+2p_2l)}\right] .$$

The total matrix element in first approximation then is the sum of all the M_1 - terms we have written down; it still is infrared divergent. The corresponding, no longer divergent matrix element of this order is according to (3,11)

$$\bar{M}_1 = M_1 + \tfrac{1}{2} B M_0 \quad .$$

and in this expression the divergent parts cancel; finally there remains

$$\bar{M}_1 = \frac{\alpha}{2\pi}\left(\frac{13}{6}\ln\frac{|q^2|}{m^2}-\frac{46}{9}\right)M_0 . \qquad (4,24)$$

This example thus illustrates equation (3,11).

2. Inelastic Electron-Scattering

We now discuss an example of inelastic electron-scattering with observation of the charged outgoing particles in coincidence. Such processes are e.g. inelastic electron-deuteron-scattering, which should give information on the nucleon form factors, or electron-pion-production. We choose the process $e + p \to e + \pi^+ + n$ (π^+ production); here radiative corrections are of importance since both electron and pion are relatively light particles which will give large contributions.

a) Experimental Situation

A sharply focussed mono-energetic electron beam (1) impinges on a proton target (2). The outgoing electron (3) will be observed at an angle Θ_{13}, the pion (4) is measured at an angle Θ_{q4} relative to the direction of momentum transfer $\vec{q} = \vec{p}_1 - \vec{p}_3$; the neutron (5) is ignored. In addition to these angles also energy (resp. momentum) of electron and pion are measured. The situation regarding the uncertainties here is even more complicated; there are two angular and two energy uncertainties, the dominating one of which has to be found. Therefore we have to study the relevant kinematics more closely. Our notation is shown in the formal diagram Fig. 46 and the momenta in Fig. 47

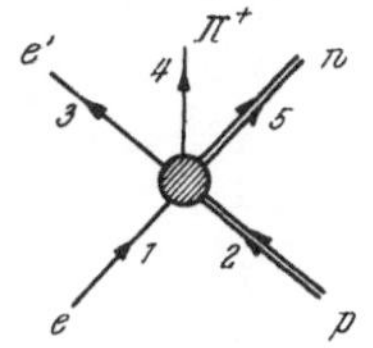

Fig. 46

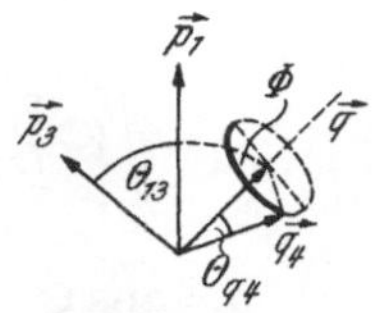

Fig. 47

Energy and momentum conservation lead to a relation somewhat more complicated than (4,4):

$F \equiv O$

$$F = \frac{M^2 + m_4^2 - m_5^2}{2} + E_4(E_3 - E_1 - M) + M(E_1 - E_3) -$$

$$- E_1 E_3 (1 - \cos \Theta_{13}) + P_4 Q \cos \Theta_{q4} \qquad (4,25)$$

where $Q = |\vec{q}|$, $P_4 = |\vec{p}_4|$. From (4,25) we shall deduce the equation for the "elastic line". Before doing this we have to consider an additional complication: in the problem the quantities E_1, E_3, E_4, Θ_{24} are not the interesting ones and therefore the experimentalists prefer other quantities. These quantities are: absolute value of the 4-momentum transfer squared (e.g. q^2, $q^2 = 1\,f^{-2}$), energy of neutron and pion in their center of mass system W (e.g. 1238 MeV corresponding to the first π-N resonance), angle between pion and $\vec{q}$ in the same c.m. system τ_4 (e.g. $\tau_4 = 90^{\circ}$). Directly we can only use Θ_{13} (e.g. 15°) and the azimuthal angle Φ. The other quantities have to be expressed by means of simple kinematical calculations in terms of the measured ones; we quote the main results:

$$q^2 \doteq - 2 E_1 E_3 (1 - \cos \Theta_{13})$$

$$W^2 = q^2 + M^2 + 2M(E_1 - E_3)$$

$$M E_{4\,(LS)} - \frac{W^2 - m_5^2 + m_4^2}{2} = Q_{(cm)} P_{4\,(cm)} \cos \tau_4 -$$

$$- \sqrt{(Q_{(cm)}^2 + q^2)(P_{4\,(cm)}^2 + m_4^2)} =$$

$$= \cos \Theta_{q4}\, Q_{(LS)} P_{4\,(LS)} -$$

$$- \sqrt{P_{4\,(LS)}^2 + m_4^2}\,(E_1 - E_3)\,. \qquad (4,26)$$

Here the quantities in the center of mass system (cm) are to be computed as follows:

$$4W^2Q^2_{(cm)} = (W^2 - q^2)^2 + M^2(M^2 - 2W^2 - 2q^2)$$

$$4W^2P^2_{4\,(cm)} = (W^2 - m_4^2)^2 + m_5^2(m_5^2 - 2W^2 - 2m_4^2) \; .$$

By means of (4,26) we can evaluate the quantities of interest for us: E_1, E_3, E_4, Θ_{q4}.

b) Uncertainties in Observation

Both for the electron counter and the pion-spectrometer we can apply the same considerations as in example 1.a). Here, however, our "measurement" lies in a 4-dimensional ($E_3 - E_4 - \Theta_{13} - \Theta_{q4}$) - space. This point is located somewhere below the "elastic plane" $dF = 0$, the plane on which the processes without radiation of additional quanta lie. The equation of this plane is (compare (4,25)):

$$a\delta E_3 + b\delta P_4 + c\delta\Theta_{13} + d\delta\Theta_{q4} = 0 \qquad (4,27)$$

with

$$a = E_4 - M - E_1(1 - \cos\Theta_{13}) + \frac{P_4}{Q}\cos\Theta_{q4}(E_3 - E_1\cos\Theta_{13})$$

$$b = (E_3 - E_1 - M)\,\frac{P_4}{E_4} + Q\cos\Theta_{q4}$$

$$c = E_1 E_3 \sin\Theta_{13}\left(\frac{P_4}{Q}\cos\Theta_{q4} - 1\right)$$

$$d = -P_4\sin\Theta_{q4}\sqrt{E_1^2 + E_3^2 - 2E_1E_3\cos\Theta_{13}} \; .$$

Since in the measurements the angular uncertainties compared to the uncertainties in energy can be neglected, as discussed in the foregoing example, we can restrict our investigation to the 2-dimensional situation in a $(E_3 - P_4)$ diagram. Electrons can be measured more exactly than pions (typical values are $\Delta E_3/E_3 = 3\ \%$, $\Delta P_4/P_4 = 10\ \%$), thus we expect a case like the one shown in Fig. 48, where obviously

$$2b\ \Delta P_4 \geq a\ \Delta E_3\ . \tag{4,28}$$

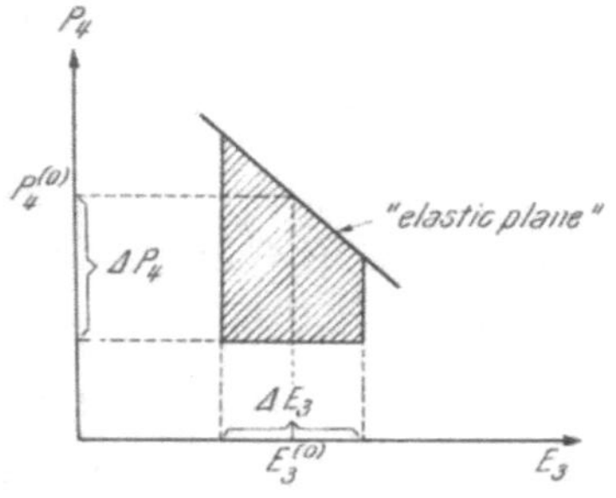

Fig. 48

In the following we assume (4,28) to be valid, i.e. ΔP_4 represents the essential uncertainty. If, however, in a particular example this should not be the case $b\,\Delta P_4$ simply has to be replaced by the actually relevant quantity (e.g. $a\,\Delta E_3$) in the final result.

We shall now investigate the influence of ΔP_4 on the radiative corrections; therefore we consider the phase-space integral

$$J = \int \frac{d^3p_5'}{E_5'} \int \frac{d^3p_3'}{E_3'} \int \frac{d^3p_4'}{E_4'} \int \frac{d^3k}{\omega}\ \delta^4(p_1 + p_2 - p_3' - p_4' - p_5' - k)\ .$$

The primed momenta are those in case of an additional photon being present. We again introduce the invariant

$$\gamma = (k \cdot p_5')$$

and compute it by comparing $(p_5' + k)^2$ with p_5^2. We get

$$\gamma = (p_1 + p_2 - p_4) \cdot \delta p_3 + (p_1 + p_2 - p_3) \cdot \delta p_4$$

which, evaluated in the lab. system, reads

$$\gamma = -(E_4 - M - E_1(1-\cos\Theta_{13}) + P_4\cos\Theta_{34})\,\delta E_3 -$$

$$-\left[(E_3 - E_1 - M)\frac{P_4}{E_4} + P_1\cos\Theta_{14} -\right.$$

$$\left. - P_3\cos\Theta_{34}\right]\delta P_4 \quad .$$

Since

$$Q\cos\Theta_{q4} = P_1\cos\Theta_{14} - P_3\cos\Theta_{34}$$

and by comparison with (4,27) we finally have

$$-\gamma = a\delta E_3 + b\delta P_4 \quad . \tag{4,29}$$

Here we explicitly assumed $\vec{p}_3$, $\vec{p}\,'_3$ and $\vec{p}_4$, $\vec{p}\,'_4$ to be parallel, i.e. sharp angular resolution. In general we therefore expect the whole expression (4,27) as the r.h.s. of (4,29).

In the integral J we perform the integrations over $d^3p'_5$ and $d\omega$ by means of the δ^4-function. The P'_4-integration we replace by one over γ and get

$$J = \int \frac{1}{E'_5}\,\frac{d^3p'_3}{E'_3}\,\frac{d^3p'_4}{E'_4}\,\frac{d^3k}{\omega}\,\delta\Big(E_1 + E_2 - \sqrt{P'^2_4 + m_4^2} -$$

$$- \sqrt{P'^2_3 + m^2} - \sqrt{k^2 + \lambda^2} -$$

$$- \sqrt{(\vec{p}_1 + \vec{p}_2 - \vec{p}\,'_3 - \vec{p}\,'_4 - \vec{k})^2 + m_5^2}\Big) =$$

$$= \int \frac{d^3p'_3}{E'_3}\,\frac{d^3p'_4}{E'_4}\,\tilde{k}^2\,\frac{\tilde{\omega}}{\tilde{k}}\,\frac{1}{\gamma}\,d\tilde{\Omega}_k\,d\tilde{E}_F\,\delta(\tilde{E}_1 + \tilde{E}_2 - \tilde{E}_F)$$

[in the special Lorentz system (~) we have

$$\tilde{\vec{p}}_1 + \tilde{\vec{p}}_2 - \tilde{\vec{p}}'_3 - \tilde{\vec{p}}'_4 = 0 \,]$$

$$J = P'_3 \, dE'_3 \, d\Omega_3 \, \frac{P'_4}{b} \, d\Omega_4 \int_0^{\Gamma} \frac{d\gamma}{\gamma} \tilde{k} \, \tilde{\omega} \, d \, \tilde{\Omega}_k \; . \qquad (4,30)$$

Here we used

$$\frac{\partial E_F}{\partial k} = \frac{\tilde{k}}{\tilde{\omega}^2} \, \frac{\gamma}{\tilde{E}'_5}$$

$$\frac{\partial \gamma}{\partial P'_4} = - b \; .$$

Again the existence of a special Lorentz system where the photon distribution is spherically symmetric was of essential importance. The limit Γ of the integration over γ is determined by

$$\Gamma = b\Delta P_4 + a\delta E_3 + c\delta\Theta_{13} + d\delta\Theta_{q4} \qquad (4,31)$$

where afterwards one has to take the average over δE_3, $\delta\Theta_{13}$, $\delta\Theta_{q4}$:

$$-\frac{\Delta E_3}{2} \leq \delta E_3 \leq + \frac{\Delta E_3}{2}$$

$$-\frac{\Delta\Theta_{13}}{2} \leq \delta\Theta_{13} \leq + \frac{\Delta\Theta_{13}}{2}$$

$$-\frac{\Delta\Theta_{q4}}{2} \leq \delta\Theta_{q4} \leq + \frac{\Delta\Theta_{q4}}{2} \; .$$

We now connect the phase-space integral (4,30) with the quantity $\tilde{B}_{ij}$ (in an analogous way as in (4,15), thereby getting the inelastic contributions). Since in paragraphs 1e, 1f and 1g we have worked with the general indices i, j . . . , we can carry over the previous results unchanged, especially the final result (4,20), but we have to consider the values of the indices i, j = 1, 2, 3, 4.

c) Evaluation of Radiative Corrections

The radiative correction δ is calculated according to (4,21) as

$$\delta = 2\alpha \sum_{ij} (\text{Re}\, B_{ij} + \tilde{B}_{ij})$$

where B_{ij} and $\tilde{B}_{ij}$ are explicitly given in (4,20). For $i \neq j$ we have to evaluate expressions of the form

$$\frac{(p_i p_j)}{\sqrt{(p_i p_j)^2 - p_i^2 p_j^2}} \{ [\ln | \frac{m_5}{2\Gamma} (\tilde{E}_i + \tilde{E}_j + (\tilde{E}_i - \tilde{E}_j) \frac{p_j^2 - p_i^2 - 2\sqrt{\ldots}}{(p_i - p_j)^2} | \times$$

$$\times \ln | \frac{2p_i^2 - 2p_i p_j - 2\sqrt{\ldots}}{2p_j^2 - 2p_i p_j + 2\sqrt{\ldots}} |] -$$

$$- [\text{ terms with the opposite sign of the square root }] +$$

$$+ [\ \Phi\text{-functions }] \} . \qquad (4,32)$$

In (4,31) we have defined Γ; we first only insert $\Gamma' = b\Delta P_4$ in (4,32) instead of Γ, thereby getting the main part of the radiative correction which we denote by δ_{soft}. The curly bracket in (4,32) can be cast into the form

$$\{ \ \} = \{ \ln \frac{P_4}{\Delta P_4} \cdot \ln | \frac{p_i^2 - p_i p_j - \sqrt{}}{p_j^2 - p_i p_j + \sqrt{}} \cdot \frac{p_j^2 - p_i p_j - \sqrt{}}{p_i^2 - p_i p_j + \sqrt{}} | +$$

$$+ \ln | \frac{m_5}{2bP_4} [\tilde{E}_i + \tilde{E}_j + (\tilde{E}_i - \tilde{E}_j) \frac{p_j^2 - p_i^2 + 2\sqrt{}}{(p_i - p_j)^2}] | \cdot$$

$$\cdot \ln | \frac{p_i^2 - p_i p_j - \sqrt{}}{p_j^2 - p_i p_j + \sqrt{}} | +$$

$$+ \ln \left| \frac{m_5}{2bP_4} \left[\tilde{E}_i + \tilde{E}_j + (\tilde{E}_i - \tilde{E}_j) \frac{p_j^2 - p_i^2 - 2\gamma}{(p_i - p_j)^2} \right] \right| \cdot$$

$$\cdot \ln \left| \frac{p_j^2 - p_i p_j - \gamma}{p_i^2 - p_i p_j + \gamma} \right| + \Phi\text{-functions} \Big\} \; . \qquad (4,33)$$

We still have to add the contributions with $i = j$ which are of much simpler form

$$\ln \frac{2\Gamma'}{m_i m_5} + \frac{\tilde{E}_i}{2\tilde{P}_i} \ln \frac{\tilde{E}_i - \tilde{P}_i}{\tilde{E}_i + \tilde{P}_i} \; . \qquad (4,33')$$

The quantities $\tilde{E}$, $\tilde{P}$ are characterized as belonging to the special Lorentz system. In (4,33) no such simple approximations can be made as in case of elastic e–p scattering. Despite this it is instructive to discuss some possible simplifications. Let us consider the case $i = 1$, $j = 3$, i.e. the electron contribution.
The first term in (4,33) is

$$2 \ln \frac{bP_4}{\Gamma'} \ln \frac{|q^2|}{m^2}$$

where we possible neglected terms of order $\sim m^2$. By means of this example we easily see how to handle the terms in Γ not taken into account up to now. We first write down the whole expression, then extract the contribution δ_{soft} and denote the remainder by δ_{hard}

$$- 2 \ln \frac{|q^2|}{m^2} \ln \frac{\Gamma}{bP_4} = - 2 \ln \frac{|q^2|}{m^2} \Big\{ \ln \frac{\Delta P_4}{P_4} +$$

$$+ \left[\ln \frac{b\Delta P_4 + a\delta E_3 + c\delta\Theta_{13} + d\delta\Theta_{q4}}{bP_4} - \right.$$

$$\left. - \ln \frac{\Delta P_4}{P_4} \right] = - 2 \ln \frac{|q^2|}{m^2} \Big\{ \ln \frac{\Delta P_4}{P_4} +$$

$$+ \left[\ln \frac{b\Delta P_4 + a\delta E_3 + c\delta\Theta_{13} + d\delta\Theta_{q4}}{b\Delta P_4} \right] \} . \qquad (4,34)$$

The second term belongs to δ_{hard}; for this expression we now have to take the average. The result of the tedious but straightforward averaging procedure is

$$M = \frac{1}{\Delta E_3} \frac{1}{\Delta\Theta_{13}} \frac{1}{\Delta\Theta_{q4}} \int d\delta\Theta_{13} \int d\delta\Theta_{q4} \int d\delta E_3 \times$$

$$\times \left[\ln \frac{b\Delta P_4 + a\delta E_3 + \dots}{b\Delta P_4} \right] = \frac{1}{6} \left\{ \frac{1}{a\Delta E_3 \, c\Delta\Theta_{13} \, d\Delta\Theta_{q4}} \times \right.$$

$$\times \left[(A + B + C + D)^3 \ln \frac{A + B + C + D}{b\Delta P_4} + \right.$$

$$\left. + \text{ 7 terms with changed signs of A, C, D } \right] - 11 \Big\}$$

(4,35)

where

$$A = a\frac{\Delta E_3}{2}, \quad B = b\Delta P_4, \quad C = c\frac{\Delta\Theta_{13}}{2}, \quad D = d\frac{\Delta\Theta_{q4}}{2} .$$

This result of course is independent of the simplified factor $- 2 \ln (|q^2| / m^2)$ in (4,34).

Altogether the contributions of δ_{hard} are in analogy to (4,33) and (4,33')

for $i \neq j$

$$- M \cdot \left\{ \ln \left| \frac{p_i^2 - p_i p_j - V}{p_j^2 - p_i p_j + V} \cdot \frac{p_j^2 - p_i p_j - V}{p_i^2 - p_i p_j + V} \right| \right\} \qquad (4,36)$$

for $i = j$

$$+ M . \qquad (4,36')$$

We now list the results of the foregoing discussion:

$$d\sigma = d\sigma_0\,(1+\delta)\ .$$

Here we put δ_{soft} into the exponential while δ_{hard} and a contribution of the vacuum polarization and the vertex

$$\delta_{el} = \frac{13}{6}\ln\frac{|q^2|}{m^2} - \frac{46}{9}\ ,$$

known already from case 1, remain as they are:

$$\delta = e^{\alpha/\pi\,\delta_{soft}}\left[1 + \frac{\alpha}{\pi}(\delta_{el} + \delta_{hard})\right] - 1$$

$$\delta_{soft} = -\sum_{i=1}^{4}(4{,}33') +$$

$$+\sum_{i=1}^{4}\sum_{i=1}^{1-1} Z_i Z_j \epsilon_i \epsilon_j \frac{(p_i p_j)}{\sqrt{(p_i p_j)^2 - p_i^2 p_j^2}}\,(4{,}33)$$

$$\delta_{hard} = -4.\,(4{,}36') + \sum_{i=1}^{4}\sum_{j=1}^{i-1} Z_i Z_j \epsilon_i \epsilon_j \frac{p_i p_j}{\sqrt{}}\,(4{,}36)\ . \qquad (4{,}37)$$

This expression (4,37) has been programmed for numerical evaluation of δ. As a typical example we show in Fig. 49 values of δ as a function of the momentum transfer.

An interesting phenomenon here is the discontinuity at $q^2 \simeq -6\,f^{-2}$. It results from the fact that there ΔE_3 and ΔP_4 interchange their roles in the determination of the essential uncertainty. Thereby δ_{hard} shows a definite maximum and so the slope of δ becomes smaller for increasing q^2. These detailed investigations, however, are beyond the scope of this survey.

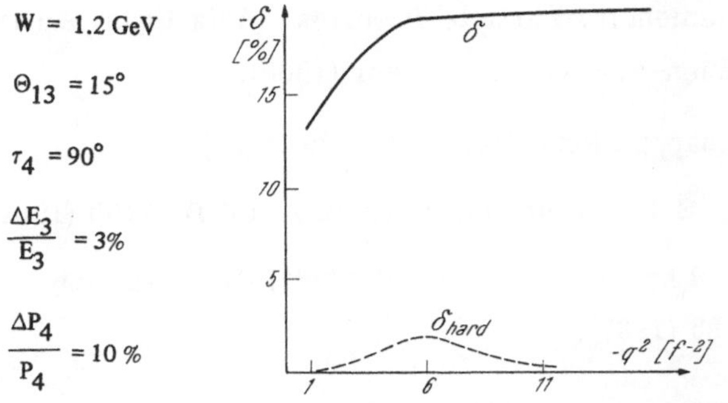

Fig. 49. Radiative corrections for $e\pi^+$ production

With these two examples we have illustrated the theoretical discussions of the previous chapter and have shown the practical application of the theory of radiative corrections to high-energy electron scattering experiments.

Literature

[1] F. Bloch, A. Nordsieck, Phys.Rev. 52, 54 (1937).

[2] W. Pauli, M. Fierz, Nuovo Cim. 15, 167 (1938).

[3] J.M. Jauch, F. Rohrlich, Helv.Phys. Acta 27, 613 (1954).

[4] J. Schwinger, Phys.Rev. 76, 780 (1949).

[5] D.R. Yennie, S.C. Frautschi, H. Suura, Ann. Phys. 13, 379 (1961).

[6] E.L. Lomon, Phys.Rev. 113, 726 (1959).

[7] R. Perrin, E.L. Lomon, Ann. Phys. 33, 328 (1965).

[8] Etim-Etim, G. Pancheri, B. Touschek, Nuovo Cim. B 51, 276 (1967).

[9] N. Meister, D.R. Yennie, Phys.Rev. 130, 1210 (1963).

[10] Y.S. Tsai, Phys.Rev. 122, 1898 (1960).

[11] A. Bartl, P. Urban, Acta Phys. Austr. 24, 87 (1966).

[12] G. Källen, Elementary Particle Theories. (Acta Phys. Austr. Suppl. III.) Wien-New York: Springer (1966).

[13] K.T. Mahanthappa, Phys.Rev. 126, 329 (1962).

[14] R.F. Dashen, S.C. Frautschi, Phys.Rev. 135 B, 1190 (1964).

[15] D.R. Yennie, Phys.Rev. 134, B 436 (1964); B. Huld, Phys. Lett. 24 B, 185 (1967).

Appendix A

THE GREEN-FUNCTIONS OF THE KLEIN-GORDON EQUATION AND THE DIRAC EQUATION

The purpose of this appendix is not to give a comprehensive description of Green functions but rather to list the various Green functions of the KG equation and the Dirac equation with some of their properties included.

The Green function $G(x, y)$ of the KG equation is defined by

$$(\Box(x) - m^2)\, G(x,y) = -\delta(x-y)$$
$$(\Box(y) - m^2)\, G(x,y) = -\delta(x-y) \quad . \tag{A,1}$$

To evaluate $G(x,y)$ we go over into the momentum space by means of a Fourier transformation and obtain after some manipulations the representation

$$G(x,y) = \frac{1}{(2\pi)^4} \int \frac{e^{-ip(x-y)}}{m^2 - p^2}\, d^4p \quad . \tag{A,2}$$

To perform the integration over p_0 we must pay attention to the poles at

$$p_0 = \pm \sqrt{\vec{p}^{\,2} + m^2}$$

and the question of how to encircle them. This question can be solved by taking into account the various boundary conditions which distinguish the special solutions of (A, 2), for these boundary conditions enable us to close the integration-contour at infinity in one way or another.

In the following we put y equal to zero.

a) The function Δ_R is defined by the boundary conditions

$$\Delta_R(x) = \frac{\partial \Delta_R(x)}{\partial x_0} = 0 \quad \text{for } x_0 < 0 \qquad (A,3)$$

which guarantee that the source at $x_0 = 0$ only acts in the future. To satisfy these conditions we have to choose the following contour:

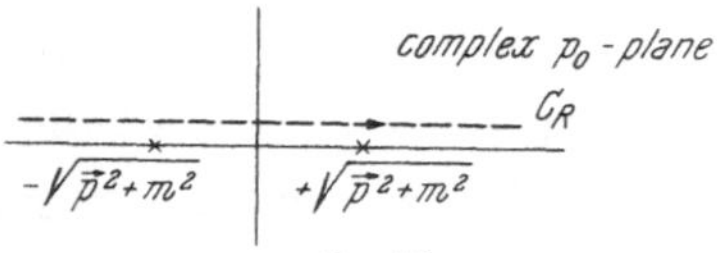

Fig. 50

For $x_0 < 0$ the contour can be closed in the upper complex p_0-half plane. In this region we have no poles. Therefore $\Delta_R(x) = 0$ in accordance with (A,3).

b) The time-reversed function is denoted by $\Delta_A(x)$ and has the boundary conditions

$$\Delta_A(x) = \frac{\partial \Delta_A(x)}{\partial x_0} = 0 \text{ for } x_0 > 0 \ . \qquad (A,4)$$

The corresponding integration contour is shown in Fig. 51.

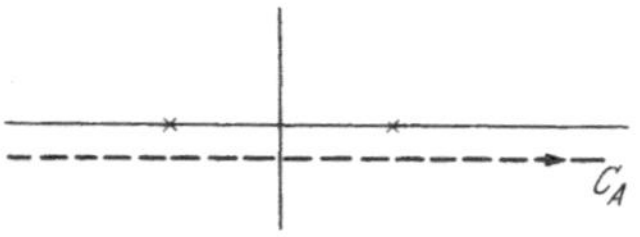

Fig. 51

c) The functions $\Delta_R^{\pm}$ are obtained from Δ_R by taking only parts of positive (negative) frequency

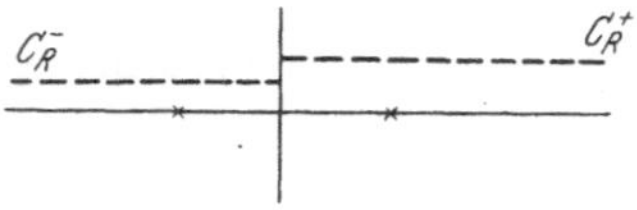

Fig. 52

and satisfy the equation

$$(\Box - m^2)\,\Delta_R^{\pm}(x) = -\,\delta(\vec{x})\,\delta_{\pm}(x_o)\ . \qquad (A,5)$$

d) In analogous manner we get the function $\Delta_A^{\pm}$

Fig. 53

e) The function Δ_+ is defined by

$$\Delta_+ = \Delta_R^+ + \Delta_A^-\ . \qquad (A,6)$$

Therefore we have to integrate over the contour

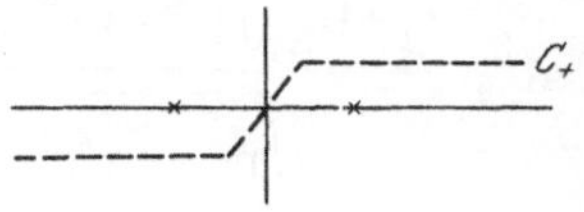

Fig. 54

This function describes the propagation of an excitation, the positive frequencies of which propagate in the future, the negative frequencies in the past (Feynman's frequency-condition). The integral-representation reads

$$\Delta_+(x) = -\frac{2\pi i}{(2\pi)^4}\int d^3p\,\frac{e^{ix_o\sqrt{\vec{p}^{\,2}+m^2}}\,e^{i\vec{p}\vec{x}}}{2\sqrt{\vec{p}^{\,2}+m^2}} \qquad x_o < 0 \qquad (A,7)$$

$$\Delta_+(x) = \frac{2\pi i}{(2\pi)^4}\int d^3p\,\frac{e^{-ix_o\sqrt{\vec{p}^{\,2}+m^2}}\,e^{i\vec{p}\vec{x}}}{2\sqrt{\vec{p}^{\,2}+m^2}} \qquad x_o > 0\ . \qquad (A,8)$$

f) Integrating over a finite closed contour we get functions which are solutions of the KG equation, for we have

$$(\Box - m^2) \oint \frac{e^{-ipx}}{m^2 - p^2} d^4p = - \oint e^{-ipx} \, d^4p \; . \qquad \text{(A,9)}$$

Since the exponential-function has no singularities except at infinity the last integral reduces to zero.

There are four non-trivial possibilities to take a finite contour

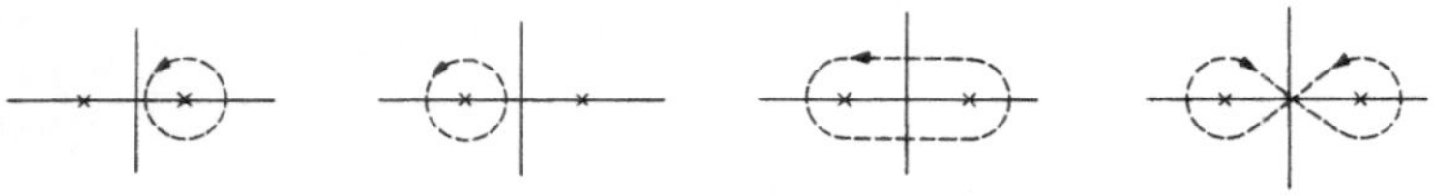

Fig. 55

The corresponding functions are denoted by $\Delta^+(x)$, $\Delta^-(x)$, $\Delta(x)$ and $\Delta^{(1)}(x)$. For $\Delta^\pm(x)$ we get the representation

$$\Delta^\pm(x) = \mp \frac{\pi i}{(2\pi)^4} \int d^3p \frac{e^{\pm i x_0 \sqrt{\vec{p}^2 + m^2}} \, e^{i \vec{p}\vec{x}}}{\sqrt{\vec{p}^2 + m^2}} \qquad \text{(A,10)}$$

with the boundary conditions

$$\frac{\partial}{\partial x_0} \Delta^\pm(x) \Big|_{x_0 = 0} = - \tfrac{1}{2} \delta(\vec{x}) \qquad \text{(A,11)}$$

and

$$\Delta^+(x) \Big|_{x_0 = 0} = - \Delta^-(x) \Big|_{x_0 = 0} \; . \qquad \text{(A,11')}$$

The function $\Delta(x)$ is connected with $\Delta^+(x)$ and $\Delta^-(x)$ by the formula

$$\Delta(x) = \Delta^+(x) + \Delta^-(x) \; . \qquad \text{(A,12)}$$

It satisfies the relations

$$(\Box - m^2) \, \Delta(x) = 0$$

$$\frac{\partial \Delta(x)}{\partial x_0} \Big|_{x_0 = 0} = - \delta(\vec{x}) \qquad \text{(A,13)}$$

$\Delta(x) = 0$ for $x^2 < 0$.

$\Delta(x)$ is useful in describing relativistic invariant commutation-relations.

Last not least we have the function

$$\Delta^{(1)}(x) = i\,(\Delta^{+}(x) - \Delta^{-}(x)) \tag{A,14}$$

with the boundary condition

$$\frac{\partial \Delta^{(1)}(x)}{\partial x_0} \Big|_{x_0 = 0} = 0 \ . \tag{A,15}$$

In the massless case we obtain the various Green functions and solutions of the equation

$$\Box \varphi = 0 \tag{A,16}$$

by means of the prescription

$$\Delta(x) \Big|_{m=0} = D(x) \ . \tag{A,17}$$

In order to get the Green functions of the Dirac equation we remember that (see appendix C)

$$(\Box - m^2)\, F(x) = +\,(i\gamma_\nu \frac{\partial}{\partial x_\nu} - m)\,(i\gamma_\mu \frac{\partial}{\partial x_\mu} + m)\, F(x) =$$

$$= -\,\delta(x) \ .$$

Defining a new function by

$$G(x) = -\,(i\gamma_\mu \frac{\partial}{\partial x_\mu} + m)\, F(x) \tag{A,18}$$

we find

$$-\,(i\gamma_\mu \frac{\partial}{\partial x_\mu} - m)\, G(x) = -\,\delta(x) \ . \tag{A,19}$$

Now $F(x)$ corresponds to the various Δ-functions enumerated above. To obtain the Green functions of the Dirac equation we only have to apply the operator

$$-\left(i\gamma_\nu \frac{\partial}{\partial x_\nu} + m\right) \tag{A,20}$$

on Δ, $\Delta^{\pm}$, Δ_{+}, Δ_{A}, Δ_{R}, $\Delta_{A}^{\pm}$, $\Delta_{R}^{\pm}$, $\Delta^{(1)}$ and get S, $S^{\pm}$, S_{+}, S_{A}, S_{R}, $S_{A}^{\pm}$, $S_{R}^{\pm}$, $S^{(1)}$ with the same integration-contours as have been used in the evaluation of the Δ's.

Appendix B

THEORY OF BOSONS, KLEIN - GORDON EQUATION

1. Derivation of the Klein-Gordon Equation

Our starting point is the Lorentz-invariant form

$$p^2 = p_\mu p_\nu g^{\mu\nu} = p_o^2 - \vec{p}^{\,2} = m^2 \qquad \text{(B,1)}$$

where p is the usual energy-momentum four-vector and m the rest-mass of the particle. In our notation we use $\hbar = c = 1$ and

$$g^{\mu\nu} = \begin{pmatrix} 1 & 0 & 0 & 0 \\ 0 & -1 & 0 & 0 \\ 0 & 0 & -1 & 0 \\ 0 & 0 & 0 & -1 \end{pmatrix}$$

In first quantization p_μ is an operator for which in the case of a free field the substitution

$$p^\mu \to i \frac{\partial}{\partial x_\mu} \equiv i\,\partial^\mu \qquad \text{(B,2)}$$

is valid.

Using the abbreviation

$$\Box \equiv - \frac{\partial^2}{\partial x_\mu \partial x^\mu} = - \frac{\partial^2}{\partial t^2} + \Delta \qquad \text{(B,3)}$$

we get the operator-equation

$$\Box - m^2 = 0$$

which operating on the wave-function $\chi(x)$ gives the KG equation

$$(\Box - m^2)\,\chi(x) = 0 \quad . \tag{B,4}$$

In the presence of an electromagnetic field the substitution (B,2) turns to be

$$p^{\mu} \rightarrow i(\partial^{\mu} + ieA^{\mu}) \equiv iD^{\mu} \quad .$$

In this case the KG equation reads

$$(-D_{\mu}D^{\mu} - m^2)\,\chi(x) = 0$$

or explicitly

$$\{-(\partial_{\mu} + ieA_{\mu})(\partial^{\mu} + ieA^{\mu}) - m^2\}\,\chi(x) = 0 \quad . \tag{B,5}$$

2. Scalar Wave Functions

We assume that the wave functions $\chi_{\mu}(x)$ are solutions of the KG equation

$$(\Box - m^2)\,\chi_k(x) = 0 \quad . \tag{B,6}$$

These solutions form a linear vector space which can be made a Hilbert-space by defining a suitable scalar product. In order to make sense this product has to be Lorentz-invariant. In addition it must not depend on time (or, more general, on a spacelike surface). These conditions may be achieved when the integrand, like a current, satisfies a continuity-equation. Thus we shall take for the integrand the form

$$J_{\mu} = \bar{\chi}_k(x) \cdot \partial_{\mu}\chi_{k'}(x) - \partial_{\mu}\bar{\chi}_k(x) \cdot \chi_{k'}(x) \equiv$$

$$\equiv \bar{\chi}_k(x)\,\overleftrightarrow{\partial}_{\mu}\,\chi_{k'}(x) \quad . \tag{B,7}$$

By applying equation (B,6) we can show that

$$\partial^{\mu} J_{\mu} = 0 \ .$$

Therefore the expression

$$\int_{\Sigma} J_{\mu} \, d\sigma^{\mu}$$

is independent of the surface Σ as long as Σ represents a space-like surface. Without restriction we can take the surface $t = \text{const}$.

$$(\chi_k, \chi_{k'}) \equiv i \int \bar{\chi}_k(x) \overleftrightarrow{\partial}_{\mu} \chi_{k'}(x) \, d\sigma^{\mu} =$$

$$= i \int \bar{\chi}_k(x) \overleftrightarrow{\partial}_{0} \chi_{k'}(x) \, d^3x \ . \qquad \text{(B,8)}$$

This scalar product has the following properties:

$$\overline{(\chi_k, \chi_{k'})} = (\chi_{k'}, \chi_k)$$

$$(\chi_k, \lambda \chi_{k'}) = \lambda (\chi_k, \chi_{k'}) \qquad \text{(B,9)}$$

with λ representing an arbitrary complex c-number. By means of a Fourier-decomposition we now rewrite equation (B,8) in the momentum space:

$$\chi_k(x) = (2\pi)^{-3/2} \int d^4p \, e^{-ipx} \, \delta(p^2 - m^2) \, \tilde{\chi}_k(p) \ . \qquad \text{(B,10)}$$

For the δ-function constraint, which tells us not to forget equation (B,1), we use the well-known formula

$$\delta(p^2 - m^2) = \frac{1}{2p^0} \{ \delta(p^0 - E_p) + \delta(p^0 + E_p) \} \ .$$

with

$$E_p = + \sqrt{\vec{p}^{\,2} + m^2} \, , \quad e^{-ipx} = e^{-ip^0 x^0 + i(\vec{p}\vec{x})} \, ,$$

$$p^0 = \pm E_p \; ; \; x^0 = t,$$

and obtain

$$\chi_k(x) = (2\pi)^{-3/2} \int \frac{d^3p}{2E_p} \{ \tilde{\chi}_k(p) e^{-ipx} + \tilde{\chi}_k(-p) e^{ipx} \} \equiv$$

$$\equiv \chi_k^+(x) + \chi_k^-(x) \; . \qquad (B,11)$$

(B,11) indicates that the linear space of functions $\chi_k(x)$ decomposes into two linear subspaces with positive and negative energy respectively. In the following we shall consider the solutions to positive energy for definiteness. The negative-energy solutions are treated in an analogous manner by simply changing (+) into (−) in the respective terms. Thus we have

$$\chi_k^+(x) = (2\pi)^{-3/2} \int \frac{d^3p}{2E_p} h_k^+(p) e^{-ipx} \qquad (B,12)$$

with

$$h_k^+(p) = \tilde{\chi}_k(p, E_p) \; .$$

In momentum space we define the scalar product to be

$$(h_k^+(p), h_{k'}^+(p)) = \int \frac{d^3p}{2E_p} \bar{h}_k^+(p) h_{k'}^+(p) \; . \qquad (B,13)$$

Inserting (B,12) into (B,8) we immediately get

$$(h_k^+(p), h_{k'}^+(p)) = (\chi_k^+(x), \chi_{k'}^+(x)) \; . \qquad (B,14)$$

It is well-known that a Hermitian operator has real eigenvalues. To each eigenvalue there exists a complete set of orthogonal eigenfunctions. In our case the relevant operator

$$(\Box - m^2)$$

is real. Therefore our functions $\chi_k^+(x)$ are orthogonal:

$$(\chi_k^+(x), \chi_{k'}^+(x)) = \delta_{k,k'} \quad . \qquad \text{(B,15)}$$

Our next aim is to exhibit a suitable completeness-relation for our orthogonal system. For this purpose we claim some mathematical ideas and definitions.

A system of orthogonal functions

$$f_1(x), f_2(x), \ldots\ldots. f_k(x)$$

is called complete in the region D if it is possible to find a number n so that if F (x) is a continuous function in D and ϵ a positive quantity,

$$\int_D | F(x) - \sum_{\nu=1}^{n} c_\nu f_\nu(x) |^2 dx < \epsilon \quad .$$

Every function F (x) for which the integrals

$$\int_D | F(x) | dx \quad \text{and} \quad \int_D | F(x) |^2 dx$$

exist can be approximated by a partial sum $F_n(x)$. If no $F_n(x)$ can be found (except probably $F_n(x) = 0$) which is orthogonal to all $f_k(x)$ we call the orthogonal system closed.

We expand two arbitrary functions F (x) and G (x) with the properties above)

$$F(x) = \sum_k c_k f_k(x) \, , \quad G(x) = \sum_k d_k f_k(x) \quad .$$

We evaluate the coefficients c_k and d_k with the help of the orthogonality-relation

$$(f_k, f_{k'}) = \delta_{k,k'}$$

and get

$$c_k = (f_k, F)\ ; \quad d_k = (f_k, G) \quad .$$

For the scalar product (F, G) we obtain

$$(F, G) = \sum_{k,k'} ((f_k, F)\, f_k, (f_{k'}, G) f_{k'}) =$$

$$= \sum_{k,k'} \overline{(f_k, F)}\, (f_{k'}, G)\, (f_k, f_{k'})$$

$$(F, G) = \sum_k (F, f_k)\, (f_k, G) \qquad \text{(B,16)}$$

which is the desired completeness-relation.

Now we choose a special F and G (see appendix A):

$$F = -\Delta^-(x-y)\ , \quad G = -\Delta^+(y-z)\ , \quad f_k = \chi_k^+(y) \quad .$$

We have

$$(F, f_k) = (-\Delta^-(x-y),\ \chi_k^+(y)) = -i \int \overline{\Delta^-(x-y)} \frac{\overleftrightarrow{\partial}}{\partial y_0} \chi_k^+(y)\, d^3y =$$

$$= i\, \chi_k^+(x) \qquad \text{(B,17)}$$

$$(f_k, G) = (\chi_k^+(y),\ -\Delta^+(y-z)) = -i \int \bar{\chi}_k^+(y) \frac{\overleftrightarrow{\partial}}{\partial y_0} \Delta^+(y-z)\, d^3y =$$

$$= i\, \bar{\chi}_k^+(z) \qquad \text{(B,18)}$$

$$(F, G) = (-\Delta^-(x-y),\ -\Delta^+(y-z)) =$$

$$= i \int \overline{\Delta^-(x-y)} \frac{\overleftrightarrow{\partial}}{\partial y_0} \Delta^+(y-z)\, d^3y = -i\,\Delta^+(x-z) \quad .$$

$$\text{(B,19)}$$

Inserting (B,17), (B,18) and (B,19) into (B,16) we finally get

$$\sum_k \chi_k^+(x)\,\bar{\chi}_k^+(z) = i\Delta^+(x-z) = \frac{1}{(2\pi)^3}\int \frac{d^3p}{2E_p}\, e^{-ip(x-z)} \quad . \qquad (B,20)$$

An analogous equation is valid for negative-energy solutions:

$$\sum_k \chi_k^-(x)\,\bar{\chi}_k^-(z) = -\,i\Delta^-(x-z) = \frac{1}{(2\pi)^3}\int \frac{d^3p}{2E_p}\, e^{ip(x-z)} \quad . \qquad (B,21)$$

In the case of plane-wave solutions we have, using a box normalization,

$$f_{\vec{p}}^{\pm}(x) = \frac{1}{\sqrt{V}}\, e^{\mp ipx} \quad .$$

For continuous eigenvalues we get

$$f_{\vec{p}}^{\pm}(x) = (2\pi)^{-3/2}\, e^{\mp ipx} \quad .$$

The scalar product now reads

$$(f_{\vec{p}}^{\pm}(x),\, f_{\vec{p}'}^{\pm}(x)) =$$

$$= \frac{i}{(2\pi)^3}\int e^{\pm ipx}\, \frac{\overleftrightarrow{\partial}}{\partial x_0}\, e^{\mp ipx}\, d^3x =$$

$$= \pm\, 2E_p\, \delta(\vec{p}-\vec{p}\,') \quad . \qquad (B,22)$$

3. Wave Functions for Particles with Spin 1

Once more we start by writing down the KG equation

$$(\Box - m^2)\,\chi_{k,\mu}(x) = 0 \quad . \qquad (B,23)$$

To obtain this equation for vector-particles by means of a Lagrangian formalism it is necessary to demand the subsidiary condition

$$\partial_\mu \chi_k^\mu(x) = 0 \quad . \qquad (B,24)$$

Just as in the previous section we use a current

$$J_\mu \equiv \bar{\chi}_k^\rho \overleftrightarrow{\partial}_\mu \chi_{k',\rho} + \partial^\rho \bar{\chi}_{k,\mu} \chi_{k',\rho} - \bar{\chi}_k^\rho \partial_\rho \chi_{k',\mu}$$

to define the scalar product

$$(\chi_{k'}^\mu, \chi_{k',\mu}) \equiv - i \int J_\mu d\sigma^\mu \quad . \tag{B,25}$$

Evaluating the integral (B,25) we see by a partial integration that the second and third term do not contribute anything because of (B,24) and the physical requirement that $\chi_k^\mu(x)$ vanish at infinity (Gauß theorem).

Therefore we get

$$(\chi_{k'}^\mu, \chi_{k',\mu}) = - i \int \bar{\chi}_k^\rho \overleftrightarrow{\partial}_\mu \chi_{k',\rho} \, d\sigma^\mu \tag{B,26}$$

(the minus sign comes from our metric) which reduces to the usual scalar product in the momentum space.

To derive the completeness-relation we introduce the operator

$$d_{\mu\nu}(x) \equiv (g_{\mu\nu} + \frac{1}{m^2} \frac{\partial^2}{\partial x^\mu \partial x^\nu}) . \tag{B,27}$$

Starting from equation (B,16) we get with

$$F = d_{\mu\nu}(x) \bar{\Delta}(x-y) , \quad G = d_{\rho\sigma}(z) \Delta^+(y-z) ,$$

$$f_k = \chi_{k,\mu}^+(y)$$

and

$$(F, f_k) = i d_{\mu\nu}(x) \chi_k^{+\nu}(x)$$

$$(f_{k'}, G) = i d_{\rho\sigma}(z) \bar{\chi}_k^{+\rho}(z) \tag{B,28}$$

$$(F, G) = i d_{\mu\nu}(x) \; d^\nu{}_\sigma(z) \Delta^+(x-z)$$

the relation

$$-\sum d_{\mu\nu}(x)\, \chi_k^{+\nu}(x)\, d_{\rho\sigma}(z)\, \bar{\chi}_k^{+\rho}(z) =$$

$$= i d_{\mu\nu}(x)\ \ d^{\nu}_{\ \sigma}(z)\, \Delta^+(x-z) \quad . \qquad \text{(B,29)}$$

To simplify this expression we first of all notice that

$$d_{\mu\nu}(x)\, \chi_k^{+\nu}(x) = g_{\mu\nu}\, \chi_k^{+\nu}(x) = \chi^+_{k,\mu}(x) \ .$$

For the righthand side of equation (B,29) we get

$$d_{\mu\nu}(x)\ \ d^{\nu}_{\ \sigma}(z)\, \Delta^+(x-z) = d_{\mu\nu}(x)\, d^{\nu}_{\ \sigma}(x)\, \Delta^+(x-z) =$$

$$= \Big[g_{\mu\sigma} + \frac{2}{m^2} \frac{\partial^2}{\partial x^{\mu} \partial x^{\sigma}} +$$

$$+ \frac{1}{m^4} \frac{\partial^2}{\partial x^{\mu} \partial x^{\sigma}} \frac{\partial^2}{\partial x^{\nu} \partial x_{\nu}} \Big] \Delta^+(x-z) =$$

$$= \Big[g_{\mu\sigma} + \frac{1}{m^2} \frac{\partial^2}{\partial x^{\mu} \partial x^{\sigma}} \Big] \Delta^+(x-z)$$

where we made use of the fact that $\Delta^+(x-z)$ satisfies the KG equation. Now (B,29) reduces to

$$\sum_k \chi^+_{k,\mu}(x)\, \bar{\chi}^+_{k,\nu}(y) =$$

$$= -i \Big[g_{\mu\nu} + \frac{1}{m^2} \frac{\partial^2}{\partial x^{\mu} \partial x^{\nu}} \Big] \Delta^+(x-y) \quad . \qquad \text{(B,30)}$$

An analogous relation is valid for negative energies. To write down a plane-wave solution for $\chi_{k,\mu}(x)$ we need a four-vector which shall be represented by the unit-vectors

$$e^{(\lambda)}_{\mu} \ ; \quad \lambda = 0,\ 1,\ 2,\ 3 \quad .$$

These vectors describe the polarization of the vector-particle and fulfil the orthogonality- and completeness-relations

$$e_{\mu}^{(\lambda)} e^{(\lambda')\mu} = -\delta^{(\lambda)(\lambda')} [1 - 2\delta^{(o)(\lambda)}]$$

$$e_{\mu}^{(o)} e_{\nu}^{(o)} - \sum_{k=1}^{3} e_{\mu}^{(k)} e_{\nu}^{(k)} = g_{\mu\nu} \quad . \tag{B,31}$$

To make sense the polarization-vector must be related to a fixed direction. The only available direction is the direction of the four-momentum p_{μ} . Therefore we choose our unit-vectors $e_{\mu}^{(\lambda)}$ in such a way as to satisfy the additional relations

$$e_{\mu}^{(o)} = \frac{1}{m} p_{\mu} \quad ; \quad (p, e^{(k)}) = 0 \quad \text{for} \quad k = 1, 2, 3 \quad .$$

Now we have

$$\sum_{\lambda=1}^{3} e_{\mu}^{(\lambda)} e_{\nu}^{(\lambda)} = -g_{\mu\nu} + \frac{p_{\mu} p_{\nu}}{m^2} \tag{B,32}$$

and the plane-wave solution reads (with respect to a finite normalization volume V)

$$\chi^{+}_{(p,\lambda)\,\mu}(x) = e_{\mu}^{(\lambda)} \frac{e^{-ipx}}{\sqrt{V}} \tag{B,33}$$

or with continuous normalization

$$\chi^{+}_{(p,\lambda)\,\mu}(x) = e_{\mu}^{(\lambda)} \frac{e^{-ipx}}{(2\pi)^{3/2}} \quad . \tag{B,34}$$

In (B,33) and (B,34) λ takes only the values 1, 2, 3 because of our subsidiary condition

$$\partial_{\mu} \chi^{\mu}_{(p,\lambda)}(x) = 0 \quad .$$

In the massless case (Maxwell-theory) equations (B,23) and (B,24) reduce to

$$\Box A_{k,\mu}(x) = 0 \,, \quad \partial_\mu A_k^\mu(x) = 0 \quad . \tag{B,35}$$

For the scalar product we assume the definition (B,26):

$$(A_{k'}^\mu, A_{k',\mu}) = -i \int \bar{A}_k^\mu(x) \frac{\overleftrightarrow{\partial}}{\partial x_o} A_{k',\mu}(x)\, d^3x \quad . \tag{B,36}$$

To obtain the completeness-relation we have to modify equation (B,30) in the following way (see appendix A):

$$d_{\mu\nu}(x) \to g_{\mu\nu} \quad ; \quad \Delta^\pm(x) \to D^\pm(x) \quad .$$

Thus we get

$$\sum_k A_{k,\mu}(x)\, \bar{A}_{k,\nu}(y) = \frac{g_{\mu\nu}}{i} D^+(x-y) \tag{B,37}$$

where $\sum$ again means summation over the momenta and polarization respectively.

The plane-wave solutions are

$$A_{k,\mu}^{(\lambda)}(x) = e_\mu^{(\lambda)} \frac{e^{-ikx}}{\sqrt{V}}$$

or

$$A_{k,\mu}^{(\lambda)}(x) = e_\mu^{(\lambda)} \frac{e^{-ikx}}{(2\pi)^{3/2}} \tag{B,38}$$

in the continuous case.

We notice, however, that in reality only transversally polarized photons exist. Therefore only two unit-vectors $e_\mu^{(1)}$ and $e_\mu^{(2)}$ are available to characterize the transverse polarization.

We have the obvious conditions

$$k^{\mu} e_{\mu}^{(1,2)} = 0 \tag{B,39}$$

$$e_{\mu}^{(i)} e^{(k)\mu} = - \delta_{ik} \qquad i, k = 1,2 \quad . \tag{B,40}$$

These equations do not determine the unit-vectors completely. Because of $k_{\mu} k^{\mu} = 0$ (in the massless case) also

$$e_{\mu}^{(i)} + \Lambda k_{\mu} \tag{B,41}$$

is a solution of equation (B,39).

This corresponds to the possibility of performing gauge-transformations, for in the coordinate space (B,41) can be written as

$$A_{\mu} \to A_{\mu} + i \frac{\partial \Lambda (x)}{\partial x^{\mu}} \quad .$$

We see that invariance under this gauge-transformations is intimately connected with the zero-mass of the photon. For the completeness-relation we now obtain (compare with (B,32))

$$e_{\mu}^{(1)} e_{\nu}^{(1)} + e_{\mu}^{(2)} e_{\nu}^{(2)} = - g_{\mu\nu} + \Lambda k_{\mu} k_{\nu} - \\ - \frac{e_{\mu}^{(3)} k_{\nu}}{(ke^{(3)})} + \frac{e_{\nu}^{(o)} k_{\mu}}{(ke^{(o)})} \quad . \tag{B,42}$$

Finally we write down the completeness-relation in the coordinate-space where the sum is to be taken only over the momenta:

$$\sum_{k} A_{\mu}^{(i)} (x) A_{\nu}^{(j)} (y) = \\ = \frac{i}{(2\pi)^2} \int e^{-ik(x-y)} e_{\mu}^{(i)} e_{\nu}^{(j)} D^{+} (k) d^4k \quad . \tag{B,43}$$

In (B,43) we put $\Lambda = 0$, a choice which can be achieved by means of a suitable gauge-transformation.

Appendix C

THEORY OF FERMIONS, DIRAC EQUATION

1. Relativistic Wave-Equation for Fermions

All wave functions which describe free particles satisfy the KG equation

$$(\Box - m^2)\,\varphi(x) = 0 \quad . \tag{C,1}$$

This equation does not take the spin-phenomena into account. Therefore Dirac tried to get an equation for fermions with spin $\frac{1}{2}$ by means of a linearization of the KG equation. He used the ansatz

$$(\Box - m^2) = \beta\,[\,i\frac{\partial}{\partial x^o} + i\sum_{k=1}^{3} \alpha^k \frac{\partial}{\partial x^k} + \\ + \beta m\,]\,\beta[\,i\frac{\partial}{\partial x^o} + i\sum_{l=1}^{3} \alpha^l \frac{\partial}{\partial x^l} - \beta m\,] \tag{C,2}$$

and postulated the equation

$$[\,i\frac{\partial}{\partial x^o} + i\sum_k \alpha^k \frac{\partial}{\partial x^k} - \beta m\,]\,\varphi(x) = 0 \quad . \tag{C,3}$$

In order to satisfy equation (C,2) α^k and β have to obey the following relations:

$$\alpha^k \alpha^l + \alpha^l \alpha^k = 2\,\delta_{kl}$$

$$\alpha^k \beta + \beta \alpha^k = 0 \qquad \text{(C,4)}$$

$$\beta^2 = 1 \quad .$$

α^k and β are represented by 4×4 matrices. In the following we shall not use α^k and β themselves but the related quantities

$$(\gamma^\mu)_{\alpha\beta} = (\beta)_{\alpha\rho}\,(\alpha^\mu)_{\rho\beta} \qquad \mu = 0,\ 1,\ 2,\ 3$$

$$\alpha^0 = \beta \quad . \qquad \text{(C,5)}$$

These quantities satisfy the relation

$$(\gamma^\mu)_{\alpha\beta}(\gamma^\nu)_{\beta\gamma} + (\gamma^\nu)_{\alpha\beta}(\gamma^\mu)_{\beta\gamma} = 2\,g^{\mu\nu}\,\delta_{\alpha\gamma} \qquad \text{(C,6)}$$

where

$$g^{\mu\nu} = \begin{pmatrix} 1 & 0 & 0 & 0 \\ 0 & -1 & 0 & 0 \\ 0 & 0 & -1 & 0 \\ 0 & 0 & 0 & -1 \end{pmatrix} .$$

We use a representation in which the matrices have the form

$$\gamma^0 = \begin{pmatrix} 1 & 0 & 0 & 0 \\ 0 & 1 & 0 & 0 \\ 0 & 0 & -1 & 0 \\ 0 & 0 & 0 & -0 \end{pmatrix} \qquad \gamma^1 = \begin{pmatrix} 0 & 0 & 0 & 1 \\ 0 & 0 & 1 & 0 \\ 0 & -1 & 0 & 0 \\ -1 & 0 & 0 & 0 \end{pmatrix}$$

(C,7)

$$\gamma^2 = \begin{pmatrix} 0 & 0 & 0 & -i \\ 0 & 0 & i & 0 \\ 0 & i & 0 & 0 \\ -i & 0 & 0 & 0 \end{pmatrix} \qquad \gamma^3 = \begin{pmatrix} 0 & 0 & 1 & 0 \\ 0 & 0 & 0 & -1 \\ -1 & 0 & 0 & 0 \\ 0 & 1 & 0 & 0 \end{pmatrix}$$

In the course of this appendix we drop the indices of the matrices except when they are necessary for the sake of clearness.
Other quantities frequently used are

$$\sigma^{\mu\nu} = \frac{1}{2}(\gamma^\mu \gamma^\nu - \gamma^\nu \gamma^\mu)$$
$$\gamma^5 \equiv i\gamma^0 \gamma^1 \gamma^2 \gamma^3 \quad . \qquad \text{(C,8)}$$

The solutions of the Dirac equation have four components because the Dirac matrices are 4×4 matrices. On the other hand we also have four linearly independent solutions of the Dirac equation. They describe positive and negative energies and the two possibilities of spin-orientation respectively. The negative-energy solutions can be interpreted as antiparticles, that means: the absence of a particle with negative energy is equivalent to the presence of an antiparticle. With the help of the Dirac matrices we can write equation (C,3) in the form

$$(i\gamma^\mu \partial_\mu - m)\,\varphi(x) = 0 \qquad \text{(C,9)}$$

Taking the Hermitian conjugate of (C,9) we get with

$$\gamma^{\mu +} \gamma^0 = \gamma^0 \gamma^\mu \qquad \text{(C,10)}$$

the adjoint equation

$$i\partial_\mu \bar{\varphi}(x)\,\gamma^\mu + m\bar{\varphi}(x) = 0 \quad . \qquad \text{(C,11)}$$

$\bar{\varphi}(x)$ is called the adjoint wave-function defined by

$$\bar{\varphi}(x) = \varphi^+(x)\,\gamma^0 \quad . \qquad \text{(C,12)}$$

Introducing the operator $\overleftarrow{\partial}$ which acts on the left we can write for (C,11)

$$\bar{\varphi}(x)\,(i\gamma^\mu \overleftarrow{\partial}_\mu - m) = 0 \quad . \qquad \text{(C,13)}$$

The quantities $\bar{\varphi}(x)$, $\varphi(x)$ and γ^μ enable us to construct a four-vector which satisfies a conservation-law:

$$s^\mu(x) = \bar{\varphi}(x)\,\gamma^\mu\,\varphi(x)\,; \quad \partial_\mu s^\mu(x) = 0 \quad . \tag{C,14}$$

Therefore the components of this vector can be interpreted as probability-density and probability-current respectively:

$$j^\mu(x) = \epsilon s^\mu(x) \quad . \tag{C,15}$$

If we perform the substitution

$$\partial_\mu \to \partial_\mu + i\epsilon A_\mu \tag{C,16}$$

we find the Dirac equation in a field

$$[\, i\gamma^\mu(\partial_\mu + i\epsilon A_\mu) - m\,]\,\varphi(x) = 0 \tag{C,17}$$

and the adjoint equation

$$\bar{\varphi}(x)\,[\, i\overleftarrow{\partial}_\mu\gamma^\mu - \epsilon A_\mu\gamma^\mu - m\,] = 0 \quad . \tag{C,18}$$

2. Charge Conjugation

In a well-known way one can interpret the negative-energy solutions of the Dirac equation as antiparticles, the positrons. It is, therefore, interesting to investigate how the theory has to be changed if the charges of all particles are changed. To this end we define a transformation, the charge conjugation, which transforms an electron solution into a positron solution. Transformed quantities will be indicated by the index c.

Let us now deduce the form of the transformation from simple physical facts. Since the free Dirac equation does not contain the charge it will be valid for φ and for φ^c:

$$(i\gamma_\nu \partial^\nu - m)\, \varphi(x) = 0 \tag{C,19}$$

$$(i\gamma_\nu \partial^\nu - m)\, \varphi^c(x) = 0 \quad . \tag{C,20}$$

In the presence of an external field the following arguments will help to find the correct equation for φ^c. Consider a given field-distribution and the forces of this field on a particle. We may ask which field exerts the same force on the antiparticle as the original one did on the particle. This field shall be named charge-conjugated field. An example illustrates the procedure: The electric field $\vec{E}$ exerts the force $\epsilon\vec{E}$ on a particle of charge ϵ, whereas the antiparticle (charge $-\epsilon$) suffers the force $-\epsilon\vec{E}^c$ in the charge-conjugated field $\vec{E}^c$. Since the forces are to be equal $\vec{E}^c = -\vec{E}$. Furthermore:

$$\vec{E} = -\frac{\partial \vec{A}}{\partial t} - \operatorname{grad} A^0 \quad . \tag{C,21}$$

Therefore:

$$A^c_\mu = -A_\mu \quad . \tag{C,22}$$

Only if the antiparticle has the same equation of motion in the charge-conjugated field as the particle in the original field the forces will be equal. Therefore φ^c must satisfy the following equation:

$$(i\gamma_\nu \partial^\nu - \epsilon\gamma_\nu A^{c\nu} - m)\, \varphi^c(x) = 0 \tag{C,23}$$

or

$$(i\gamma_\nu \partial^\nu + \epsilon\gamma_\nu A^\nu - m)\, \varphi^c(x) = 0 \quad . \tag{C,24}$$

In order find φ^c we take the complex conjugate of the original equation:

$$(i\gamma_\nu \partial^\nu - \epsilon\gamma_\nu A^\nu - m)_{\alpha\beta}\, \varphi_\beta(x) = 0 \tag{C,25}$$

$$(-i\gamma^*_\nu \partial^\nu - \epsilon\gamma^*_\nu A^\nu - m)_{\alpha\beta}\, \varphi^*_\beta(x) = 0 \quad . \tag{C,26}$$

Postulating

$$(\gamma^*_\nu)_{\alpha\beta} = -(D^{-1})_{\alpha\gamma} (\gamma_\nu)_{\gamma\delta} D_{\delta\beta} \tag{C,27}$$

the Dirac equation reads

$$(D^{-1})_{\alpha\beta} (i\gamma_\nu \partial^\nu + \epsilon\gamma_\nu A^\nu - m)_{\gamma\delta} D_{\delta\beta}\, \varphi^*_\beta(x) = 0 \quad . \tag{C,28}$$

This corresponds exactly to equation (C,24) if one defines:

$$\varphi^c_\alpha = D_{\alpha\beta}\, \varphi^*_\beta \quad . \tag{C,29}$$

It can easily be shown that such a Dirac matrix D exists. The explicit form of D depends on the representation of the γ's, nevertheless D has very general properties. Since $(\varphi^c)^c = \varphi$,

$$(\varphi^c)^c = D\varphi^{c*} = D\,(D\varphi^*)^* = DD^*\varphi \quad . \tag{C,30}$$

Therefore

$$DD^* = 1 \tag{C,31}$$

independent of the representation.
In our γ - representation (C,7):

$$\gamma^*_0 = \gamma_0 \,, \quad \gamma^*_1 = \gamma_1 \,, \quad \gamma^*_3 = \gamma_3 \,; \quad \gamma^*_2 = -\gamma_2 \quad .$$

Using these relations and equation (C,27), the only possibility for D is

$$D = \lambda\gamma^2 \quad \text{with} \quad \lambda\lambda^* = 1 \quad . \tag{C,32}$$

We choose

$$D = i\gamma^2 \quad . \tag{C,33}$$

In this representation D is Hermitian and its square equal to 1. There exists another set of γ's for which D is a multiple of the unit-matrix, so that the charge-conjugated solution is identical to the complex conjugated one up to a factor. The transition to the new representation is given by the following matrix S:

$$\gamma^{\mu'} = S\gamma^{\mu}S^{-1} \tag{C,34}$$

$$S = \frac{1}{2} \cdot \begin{pmatrix} -1 & 1 & 1 & 1 \\ 1 & 1 & 1 & 1 \\ -i & i & -i & -i \\ i & i & -i & i \end{pmatrix} \quad . \tag{C,35}$$

Concerning the particle current s^{μ} and the electric current j^{μ} one can shown that $s^{c\mu} = s^{\mu}$. Therefore

$$j^{c\mu} = \epsilon s^{c\mu} = \epsilon s^{\mu} = j^{\mu} \quad .$$

This latter relation contradicts the physical fact that the electric current of antiparticles and particles must have opposite signs. This discrepancy can be solved in a consequent multiparticle theory (field-theory).

3. Solutions for a Free Particle

The solutions of the Dirac equation for a free particle may be written as plane waves. There are two solutions for positive and negative energy states respectively:

$$\varphi^{(1,2)}(x) = \frac{N}{\sqrt{V}} u^{(1,2)}(\vec{p}) \cdot$$

$$\cdot \, e^{-i\sqrt{\vec{p}^{\,2} + m^2}\, x_0 + i\vec{p}\vec{x}} \tag{C,36}$$

$$\varphi^{(3,4)}(x) = \frac{N}{\sqrt{V}} u^{(3,4)}(\vec{p}) \cdot$$

$$\cdot\, e^{+i\sqrt{\vec{p}^{\,2}+m^2}\,x_0 + i\vec{p}\vec{x}} \quad . \qquad \text{(C,37)}$$

N is a normalization factor and V the quantization volume. Combining these two equations we get:

$$\varphi^{(\rho)}(x) = \frac{N}{\sqrt{V}} u^{(\rho)}(\vec{p})\, e^{-ipx} \qquad \text{(C,38)}$$

(for $\rho = 1,2$; $p_0 = +\sqrt{\vec{p}^{\,2}+m^2}$ and for $\rho = 3,4$; $p_0 = -\sqrt{\vec{p}^{\,2}+m^2}$).
The spinors $u^{(\rho)}(\vec{p})$ satisfy

$$(p_\mu \gamma^\mu - m)\, u^{(\rho)}(\vec{p}) = 0 \quad . \qquad \text{(C,39)}$$

We construct linear combinations of these $u^\rho(\vec{p})$ which are eigenfunctions of the spinoperator $(\vec{\sigma}\vec{p})$:

$$(\vec{\sigma}\vec{p})\, u^{(1,3)}(\vec{p}) = |\vec{p}|\, u^{(1,3)}(\vec{p})$$

$$(\vec{\sigma}\vec{p})\, u^{(2,4)}(\vec{p}) = -|\vec{p}|\, u^{(2,4)}(\vec{p}) \quad . \qquad \text{(C,40)}$$

The positron wave functions are electron wave functions with negative energy and negative momentum. We therefore define positron solutions $v^{(1,2)}$:

$$v^{(1,2)}(\vec{p}) = u^{(3,4)}(-\vec{p}) \quad . \qquad \text{(C,41)}$$

Instead of the four solutions (C,38) we now use the following set:

$$\varphi^{(s)}(x) = \frac{N}{\sqrt{V}} u^{(s)}(\vec{p})\, e^{-ipx}$$

$$\varphi'^{(s)}(x) = \frac{N}{\sqrt{V}} v^{(s)}(\vec{p})\, e^{+ipx} \qquad s = 1,2 \quad . \qquad \text{(C,42)}$$

Since φ^s, φ'^s are solutions of Dirac's equation, the u's and v's must satisfy the equations

$$(p_\mu \gamma^\mu - m)\, u^{(s)}(\vec{p}) = 0$$

$$(p_\mu \gamma^\mu + m)\, v^{(s)}(\vec{p}) = 0 \quad . \tag{C,43}$$

Sometimes it is useful to project a spinor with definite spin orientation and definite energy from a general Dirac solution by means of so-called projection operators. The energy projection operators are

$$\Lambda_\pm(\vec{p}) = \frac{m \pm p_\mu \gamma^\mu}{2m} \quad . \tag{C,44}$$

They have the following properties

$$\Lambda_+(\vec{p})\, u^{(s)}(\vec{p}) = u^{(s)}(\vec{p})$$

$$\Lambda_-(\vec{p})\, v^{(s)}(\vec{p}) = v^{(s)}(\vec{p}) \tag{C,45}$$

$$\Lambda_+(\vec{p})\, v^{(s)}(\vec{p}) = \Lambda_-(\vec{p})\, u^{(s)}(\vec{p}) = 0 \quad .$$

In addition they have the properties of any projection operator

$$\Lambda_+ + \Lambda_- = 1$$

$$(\Lambda_\pm)^2 = \Lambda_\pm\, ; \quad \Lambda_+ \Lambda_- = \Lambda_- \Lambda_+ = 0 \quad . \tag{C,46}$$

The spin projection operators are

$$\Sigma_\pm(p) = \frac{|\vec{p}| \pm (\vec{\sigma}\vec{p})}{2|\vec{p}|} \tag{C,47}$$

with the properties

$$\Sigma_-(\vec{p})\, v^{(1)}(\vec{p}) = v^{(1)}(\vec{p})$$

$$\Sigma_+ (\vec{p})\, v^{(2)} (\vec{p}) = v^{(2)} (\vec{p})$$

$$\Sigma_+ (\vec{p})\, u^{(1)} (\vec{p}) = u^{(1)} (\vec{p})$$

$$\Sigma_- (\vec{p})\, u^{(2)} (\vec{p}) = u^{(2)} (\vec{p}) \qquad \text{(C,48)}$$

$$\Sigma_+ (\vec{p})\, v^{(1)} (\vec{p}) = \Sigma_- (\vec{p})\, v^{(2)} (\vec{p}) = 0$$

$$\Sigma_- (\vec{p})\, u^{(1)} (\vec{p}) = \Sigma_+ (\vec{p})\, u^{(2)} (\vec{p}) = 0 \quad .$$

Once more the general projection operator conditions are valid. By explicit calculations one easily finds

$$[\Lambda_\pm, \Sigma_\pm]_- = [\Lambda_\pm, \Lambda_\pm]_- = [\Sigma_\pm, \Sigma_\pm]_- = 0 \quad . \qquad \text{(C,49)}$$

Let us now consider an arbitrary spinor χ:

$$\chi (\vec{p}) = \sum_s (\lambda^{(s)} u^{(s)} (\vec{p}) + \mu^{(s)} v^{(s)} (\vec{p})) \quad . \qquad \text{(C,50)}$$

Since the projection operators commute the functions $u^{(s)}$, $v^{(s)}$ are simultaneous eigenspinors of $\Sigma_\pm$ and $\Lambda_\pm$. Therefore we get a spinor of definite spin and energy by means of the following combination of projection operators:

$$\Lambda_+ (\vec{p})\, \Sigma_+ (\vec{p})\, \chi (\vec{p}) = \lambda^{(1)} u^{(1)} (\vec{p})$$

$$\Lambda_+ (\vec{p})\, \Sigma_- (\vec{p})\, \chi (\vec{p}) = \lambda^{(2)} u^{(2)} (\vec{p})$$

$$\Lambda_- (\vec{p})\, \Sigma_- (\vec{p})\, \chi (\vec{p}) = \mu^{(1)} v^{(1)} (\vec{p}) \qquad \text{(C,51)}$$

$$\Lambda_- (\vec{p})\, \Sigma_+ (\vec{p})\, \chi (\vec{p}) = \mu^{(2)} v^{(2)} (\vec{p}) \quad .$$

One easily shows the validity of

$$\gamma_0 \Lambda_\pm^+ \gamma_0 = \Lambda_\pm ; \quad \gamma_0 \Sigma_\pm^+ \gamma_0 = \Sigma_\pm \; . \tag{C,52}$$

By using (C,52) we can derive the following orthogonality-relations:

$$\begin{aligned} \bar{u}^{(r)} u^{(s)} &= \bar{u}^{(s)} u^{(r)} = \delta_{rs} \\ \bar{v}^{(r)} v^{(s)} &= \bar{v}^{(s)} v^{(r)} = - \delta_{rs} \\ \bar{u}^{(r)} v^{(s)} &= \bar{v}^{(r)} u^{(s)} = 0 \; . \end{aligned} \tag{C,53}$$

Now we show that the normalization chosen in (C,53) is a consequence of the requirement that the normalization of the wave-functions should be Lorentz-invariant. The density of the probability current is (number of particles per unit volume)

$$s^0 = \bar{\varphi} \gamma^0 \varphi = \varphi^+ \varphi = u^+ u \frac{N^2}{V} \; . \tag{C,54}$$

The number of particles is invariant, but not the volume:

$$\begin{aligned} d^3x' &= d^3x \sqrt{1-v^2} \\ \frac{|E'|}{|E|} &= \frac{1}{\sqrt{1-v^2}} \end{aligned} \quad \rightarrow \frac{d^3x'}{d^3x} = \frac{|E|}{|E'|} \; . \tag{C,55}$$

The volume transforms like $\frac{1}{|E|}$, consequently the particle density transforms like $|E|$. In order to satisfy the invariance of the normalization factor N^2/V we have to choose

$$u^+ u = \frac{|E|}{m} \; . \tag{C,56}$$

By virtue of

$$H u = (\vec{\alpha}\vec{p} + \beta m) u = E u \; ; \quad u^+ H = E u^+$$

we find

$$u^+\{\beta, H\}u = 2Eu^+\beta u = 2E\bar{u}u \quad . \tag{C,57}$$

Furthermore

$$\{\beta, H\} = 2m \quad ,$$

$$u^+u = \frac{E}{m}\bar{u}\,u \quad .$$

This leads to our normalization (C,53) :

$$u^+u = \frac{|E|}{m} = \frac{E}{m}\bar{u}u$$

$$\bar{u}u = \frac{|E|}{E} = \pm 1 \qquad \text{for} \quad \begin{matrix} E > 0 \\ E < 0 \end{matrix} \quad . \tag{C,58}$$

Since

$$\int_V \bar{\varphi}\varphi\, d^3x = N^2 u^+u = \frac{|E|}{m} N^2 = 1$$

we get

$$N = \sqrt{\frac{m}{|E|}} \quad . \tag{C,59}$$

To obtain the explicit form of the free Dirac equation we start with (C,43), (C,48)

$$(p_\mu \gamma^\mu - m)\, u^{(1)} = \begin{pmatrix} p^0 - m & -(\vec{\sigma}\vec{p}) \\ (\vec{\sigma}\vec{p}) & -p^0 - m \end{pmatrix} \cdot \begin{pmatrix} a \\ b \end{pmatrix} = 0 \tag{C,60}$$

$$(|\vec{p}| - (\vec{\sigma}\vec{p}))\, u^{(1)} = \left[\begin{pmatrix} |\vec{p}| & 0 \\ 0 & |\vec{p}| \end{pmatrix} + \right.$$

$$+ \begin{pmatrix} -(\vec{\sigma}\vec{p}) & 0 \\ 0 & -(\vec{\sigma}\vec{p}) \end{pmatrix} \Bigg] \begin{pmatrix} a \\ b \end{pmatrix} = 0$$

(C,61)

where

$$u^{(1)} = \begin{pmatrix} a \\ b \end{pmatrix}$$

and a, b are two-component quantities.
To fulfil

$$[\, |\vec{p}\,| - (\vec{\sigma}\vec{p})\,]\, a = \begin{pmatrix} |\vec{p}\,| - p^3, & -p^- \\ -p^+ & |\vec{p}\,| + p^3 \end{pmatrix} a = 0 \qquad \text{(C,62)}$$

$$p^{\pm} = p^1 \pm ip^2$$

we put

$$a = \alpha \begin{pmatrix} |\vec{p}\,| + p^3 \\ p^+ \end{pmatrix} \qquad \text{(C,63)}$$

where α is a normalization constant.
Furthermore from

$$(p^0 - m)\, a - (\vec{\sigma}\vec{p})\, b = 0$$

$$[\, |\vec{p}\,| - (\vec{\sigma}\vec{p})\,]\, b = 0 \qquad \text{(C,64)}$$

we get

$$b = \frac{p^0 - m}{|\vec{p}\,|}\, a \qquad \text{(C,65)}$$

and with help of (C,53)

$$\bar{u}^{(1)} u^{(1)} = (a^+ b^+) \begin{pmatrix} 1 & 0 \\ 0 & -1 \end{pmatrix} \cdot \begin{pmatrix} a \\ b \end{pmatrix} =$$

$$= a^+ a - b^+ b = 1 \quad . \tag{C,66}$$

Combining (C,63), (C,65), (C,66) we get

$$\alpha^2 = \frac{|\vec{p}|}{4m\,(p^0 - m)\,(\,|\vec{p}| + p^3)} \quad . \tag{C,67}$$

Using the energy-momentum-relation

$$\frac{p^0 - m}{|\vec{p}|} = \frac{|\vec{p}|}{p^0 + m}$$

we find

$$u^{(1)} = A \begin{pmatrix} 1 \\ \dfrac{p^+}{|\vec{p}| + p^3} \\ \dfrac{p^0 - m}{|\vec{p}|} \\ \dfrac{p^+\,|\vec{p}|}{(p^0 + m)\,(\,|\vec{p}| + p^3)} \end{pmatrix} \tag{C,68}$$

where

$$A = \frac{1}{2\sqrt{m}} \cdot \sqrt{\frac{(p^0 + m)\,(\,|\vec{p}| + p^3)}{|\vec{p}|}} \; . \tag{C,69}$$

In a similar way we can derive the other spinors

$$u^{(2)} = A \begin{pmatrix} -\dfrac{p^-}{|\vec{p}| + p^3} \\ 1 \\ \dfrac{|\vec{p}|\, p^-}{(p^o + m)\,(|\vec{p}| + p^3)} \\ -\dfrac{|\vec{p}|}{p^o + m} \end{pmatrix} \qquad (C,70)$$

$$v^{(1)} = A \begin{pmatrix} \dfrac{|\vec{p}|\, p^-}{(p^o + m)\,(|\vec{p}| + p^3)} \\ -\dfrac{|\vec{p}|}{p^o + m} \\ -\dfrac{p^-}{|\vec{p}| + p^3} \\ 1 \end{pmatrix} \qquad (C,71)$$

$$v^{(2)} = A \begin{pmatrix} \dfrac{|\vec{p}|}{p^o + m} \\ \dfrac{|\vec{p}|\, p^+}{(p^o + m)\,(|\vec{p}| + p^3)} \\ 1 \\ \dfrac{p^+}{|\vec{p}| + p^3} \end{pmatrix} \qquad (C,72)$$

$u^{(r)}$, $v^{(r)}$ not only form an orthonormal but also a complete system with respect to the degrees of freedom for spin and sign of energy:

$$\sum_{r=1,2} \{ u_\alpha^{(r)}(\vec{p})\, \bar{u}_\beta^{(r)}(\vec{p}) - v_\alpha^{(r)}(\vec{p})\, \bar{v}_\beta^{(r)}(\vec{p}) = \delta_{\alpha\beta} \ .$$

$$\text{(C,73)}$$

This can be shown by

$$\sum_{r=1,2} u_\alpha^{(r)}(\vec{p})\, \bar{u}_\beta^{(r)}(\vec{p}) = (\Lambda_+)_{\alpha\beta} \qquad \text{(C,74)}$$

$$\sum_{r=1,2} v_\alpha^{(r)}(\vec{p})\, \bar{v}_\beta^{(r)}(\vec{p}) = -(\Lambda_-)_{\alpha\beta} \qquad \text{(C,75)}$$

and

$$\Lambda_+ + \Lambda_- = 1 \ .$$

Orthogonality and completeness may also be generalized to plane-wave solutions:

$$\varphi^{(\rho)(\vec{p})}(x) = \frac{1}{\sqrt{V}} \sqrt{\frac{m}{|E|}}\; u_\alpha^{(\rho)}(\vec{p})\, e^{-ipx} \qquad \text{(C,76)}$$

$$p_0 = +\sqrt{|\vec{p}|^2 + m^2} \quad \text{for } \rho = 1,2 \ ;$$

$$p_0 = -\sqrt{|\vec{p}|^2 + m^2} \quad \text{for } \rho = 3,4 \ .$$

The index $\vec{p}$ represents the additional discrete degrees of freedom. These $\varphi_\alpha^{(\rho)(\vec{p})}(x)$ are orthonormal:

$$\int_V \varphi_\alpha^{+(\sigma)(\vec{p})}(x)\,\varphi_\alpha^{(\rho)(\vec{p}\,')}(x)\,d^3x =$$

$$= \frac{m}{|E|V} \int_V d^3x\, e^{ix(p-p')}\, u_\alpha^{+(\rho)}(p) \times u_\alpha^{(\rho)}(p') =$$

$$= \frac{m}{|E|}\,\delta_{\vec{p},\vec{p}\,'}\, e^{ix_0(p_0 - p'_0)}\, u_\alpha^{+(\sigma)}(\vec{p})\, u_\alpha^{(\rho)}(\vec{p}) =$$

$$= \delta_{\vec{p},\vec{p}\,'}\,\delta_{\sigma\rho} \quad . \tag{C,77}$$

To derive (C,77) we used (C,53), (C,56).

Now we shall derive the generalized completeness-relation

$$\sum_{(\vec{p})} \sum_{\rho=1}^{4} \varphi_\alpha^{(\rho)(\vec{p})}(x)\,\bar{\varphi}_\beta^{(\rho)(\vec{p})}(x') =$$

$$= \sum_\rho \sum_{\vec{p}} \frac{m}{V|E|}\, e^{-ip(x-x')}\, u_\alpha^{(\rho)}(\vec{p})\,\bar{u}_\beta^{(\rho)}(\vec{p}) \quad . \tag{C,78}$$

Using

$$\sum_{\rho=3}^{4} u_\alpha^{(\rho)}(\vec{p})\,\bar{u}_\beta^{(\rho)}(\vec{p}) = \sum_{\rho=1}^{2} v_\alpha^{(\rho)}(-\vec{p})\,\bar{v}_\beta^{(\rho)}(-\vec{p}) =$$

$$= -\left(\Lambda_+(-\vec{p})\right)_{\alpha\beta} = -\left(\Lambda_+(\vec{p})\right)_{\alpha\beta} \tag{C,79}$$

we obtain

$$\sum_\rho e^{-ip(x-x')}\, u_\alpha^{(\rho)}(\vec{p})\,\bar{u}_\beta^{(\rho)}(\vec{p}) = \left(\Lambda_+(\vec{p})\right)_{\alpha\beta} \times$$

$$\times \{ e^{-i|p_0|(x_0 - x'_0)} -$$

$$- e^{i|p_0|(x_0 - x'_0)} \} e^{i\vec{p}(\vec{x}-\vec{x}')} \quad . \qquad (C,80)$$

We finally get

$$\sum_{\vec{p}} \sum_{\rho=1}^{4} \varphi_{\alpha}^{(\rho)(\vec{p})}(x)\, \bar{\varphi}_{\beta}^{(\rho)(\vec{p})}(x') =$$

$$= \frac{1}{V} \sum_{\vec{p}} \frac{(p_\mu \gamma^\mu + m)_{\alpha\beta}}{2|p_0|} [e^{-i|\vec{p}_0|(x_0 - x'_0)} -$$

$$- e^{+i|p_0|(x_0 - x'_0)}] e^{i\vec{p}(\vec{x}-\vec{x}')} \quad . \qquad (C,81)$$

$\frac{1}{V} \sum_{\vec{p}}$ changes to $\frac{1}{(2\pi)^3} \int d^3p$ in continuum normalization.

The S-function, defined in appendix A, reads

$$S_{\alpha\beta}(x-x') = \frac{i}{(2\pi)^3} \int d^3p \frac{(p_\mu \gamma^\mu + m)_{\alpha\beta}}{2|p_0|} [e^{-i|p_0|(x_0 - x'_0)} -$$

$$- e^{+i|p_0|(x_0 - x'_0)}] e^{i\vec{p}(\vec{x}-\vec{x}')} \quad . \qquad (C,82)$$

The generalized completeness-relation gets

$$\sum_{\vec{p}} \sum_{\rho=1}^{4} \varphi_{\alpha}^{(\rho)(\vec{p})}(x)\, \bar{\varphi}_{\beta}^{(\rho)(\vec{p})}(x') =$$

$$= i\, S_{\alpha\beta}(x-x') \quad . \qquad (C,83)$$

Finally we write down matrix elements of some important matrices sandwiched between electron and positron states. Of course, these matrix elements are numbers and not spin matrices. For simplicity we arrange them in a quadratic scheme:

Γ	$\tilde{u}_\alpha^{(r)} \Gamma_{\alpha\beta} u_\beta^{(s)}$	$\tilde{u}_\alpha^{(r)} \Gamma_{\alpha\beta} v_\beta^{(s)}$	$\tilde{v}_\alpha^{(r)} \Gamma_{\alpha\beta} u_\beta^{(s)}$	$\tilde{v}_\alpha^{(r)} \Gamma_{\alpha\beta} v_\beta^{(s)}$	
	rows: $r=1$, $r=2$; columns: $s=1$, $s=2$				
1	$\begin{matrix} 1 & 0 \\ 0 & 1 \end{matrix}$	$\begin{matrix} 0 & 0 \\ 0 & 0 \end{matrix}$	$\begin{matrix} 0 & 0 \\ 0 & 0 \end{matrix}$	$\begin{matrix} -1 & 0 \\ 0 & -1 \end{matrix}$	
γ^0	$\begin{matrix} \frac{p^0}{m} & 0 \\ 0 & \frac{p^0}{m} \end{matrix}$	$\begin{matrix} \frac{p^3}{m} & \frac{p^-}{m} \\ \frac{p^+}{m} & \frac{p^3}{m} \end{matrix}$	$\begin{matrix} \frac{p^3}{m} & \frac{p^-}{m} \\ \frac{p^+}{m} & -\frac{p^3}{m} \end{matrix}$	$\begin{matrix} \frac{p^0}{m} & 0 \\ 0 & \frac{p^0}{m} \end{matrix}$	
γ^1	$\begin{matrix} \frac{p^1}{m} & 0 \\ 0 & \frac{p^1}{m} \end{matrix}$	$\frac{1}{2m\Delta}\begin{pmatrix} 2p^1p^3, & B^- + \Delta^2 \\ B^+ + \Delta^2, & -2p^1p^3 \end{pmatrix}$	$\frac{1}{2m\Delta}\begin{pmatrix} 2p^1p^3, & B^- + \Delta^2 \\ B^+ + \Delta^2, & -2p^1p^3 \end{pmatrix}$	$\begin{matrix} \frac{p^1}{m} & 0 \\ 0 & \frac{p^1}{m} \end{matrix}$	$B^\pm = (p^\pm)^2 - (p^2)^2$
γ^2	$\begin{matrix} \frac{p^2}{m} & 0 \\ 0 & \frac{p^2}{m} \end{matrix}$	$\frac{1}{2m\Delta}\begin{pmatrix} 2p^2p^3, & i(D^- - \Delta^2) \\ -i(D^+ - \Delta^2), & -2p^2p^3 \end{pmatrix}$	$\frac{1}{2m\Delta}\begin{pmatrix} 2p^2p^3, & i(D^- - \Delta^2) \\ -i(D^+ - \Delta^2), & -2p^2p^3 \end{pmatrix}$	$\begin{matrix} \frac{p^2}{m} & 0 \\ 0 & \frac{p^2}{m} \end{matrix}$	$D^\pm = (p^\pm)^2 + (p^3)^2$
γ^3	$\begin{matrix} \frac{p^3}{m} & 0 \\ 0 & \frac{p^3}{m} \end{matrix}$	$\frac{1}{2m\Delta}\begin{pmatrix} q^2 + \Delta^2, & 2p^3p^- \\ 2p^3p^+, & -q^2 - \Delta^2 \end{pmatrix}$	$\frac{1}{2m\Delta}\begin{pmatrix} \Delta^2 + q^2, & 2p^3p^- \\ 2p^3p^+, & -\Delta^2 - q^2 \end{pmatrix}$	$\begin{matrix} \frac{p^3}{m} & 0 \\ 0 & \frac{p^3}{m} \end{matrix}$	$q^2 = (p^3)^2 - p^+p^-$ $\Delta = \lvert p^0 \rvert + m$
γ^5	$\begin{matrix} 0 & 0 \\ 0 & 0 \end{matrix}$	$\begin{matrix} 1 & 0 \\ 0 & 1 \end{matrix}$	$\begin{matrix} -1 & 0 \\ 0 & -1 \end{matrix}$	$\begin{matrix} 0 & 0 \\ 0 & 0 \end{matrix}$	
$\gamma_o(\vec{\gamma}\vec{p})$	$\begin{matrix} 0 & 0 \\ 0 & 0 \end{matrix}$	$\begin{matrix} p^3 & p^- \\ p^+ & -p^3 \end{matrix}$	$\begin{matrix} -p^3 & -p^- \\ -p^+ & p^3 \end{matrix}$	$\begin{matrix} 0 & 0 \\ 0 & 0 \end{matrix}$	

Γ	$\tilde{u}_\alpha^{(r)}\Gamma_{\alpha\beta}u_\beta^{(s)}$	$\tilde{u}_\alpha^{(r)}\Gamma_{\alpha\beta}v_\beta^{(s)}$	$\tilde{v}_\alpha^{(r)}\Gamma_{\alpha\beta}u_\beta^{(s)}$	$\tilde{v}_\alpha^{(r)}\Gamma_{\alpha\beta}v_\beta^{(s)}$
$i\gamma^1\gamma^2\gamma^3$	$\begin{matrix} \frac{p^3}{m} & \frac{p^-}{m} \\ \frac{p^+}{m} & -\frac{p^3}{m} \end{matrix}$	$\begin{matrix} \frac{p^o}{m} & 0 \\ 0 & \frac{p^0}{m} \end{matrix}$	$\begin{matrix} \frac{p^0}{m} & 0 \\ 0 & \frac{p^0}{m} \end{matrix}$	$\begin{matrix} \frac{p^3}{m} & \frac{p^-}{m} \\ \frac{p^+}{m} & -\frac{p^3}{m} \end{matrix}$
$(\vec{\sigma}\vec{p})\gamma_0$	$\frac{p_o}{m}\begin{pmatrix} p^3 & p^- \\ p^+ & -p^3 \end{pmatrix}$	$\begin{matrix} \frac{p_o^2-m^2}{m} & 0 \\ 0 & \frac{p_o^2-m^2}{m} \end{matrix}$	$\begin{matrix} \frac{p_o^2-m^2}{m} & 0 \\ 0 & \frac{p_o^2-m^2}{m} \end{matrix}$	$\frac{p^o}{m}\begin{pmatrix} p^3 & p^- \\ p^+ & -p^3 \end{pmatrix}$
σ^1	$\frac{1}{2m\Delta}\begin{pmatrix} 2p^3p^1, & \Delta^2+D^- \\ \Delta^2+B^+, & -2p^3p^1 \end{pmatrix}$	$\frac{1}{m}\begin{pmatrix} ip^2 & -p^3 \\ p^3 & -ip^2 \end{pmatrix}$	$-\frac{1}{m}\begin{pmatrix} ip^2 & -p^3 \\ p^3 & -ip^2 \end{pmatrix}$	$-\frac{1}{2m\Delta}\begin{pmatrix} 2p^3p^1, & \Delta^2+B^- \\ \Delta^2+B^+, & -2p^3p^1 \end{pmatrix}$
σ^2	$\frac{i}{2m\Delta}\begin{pmatrix} -2ip^3p^2, & D^- - \Delta^2 \\ \Delta^2-D^+, & 2ip^3p^2 \end{pmatrix}$	$\frac{i}{m}\begin{pmatrix} p^1 & -p^3 \\ -p^3 & -p^1 \end{pmatrix}$	$-\frac{i}{m}\begin{pmatrix} p^1 & -p^3 \\ -p^3 & -p^1 \end{pmatrix}$	$-\frac{i}{2m\Delta}\begin{pmatrix} -2ip^3p^2, & -\Delta^2+D^- \\ \Delta^2-D^+, & 2ip^3p^2 \end{pmatrix}$
σ^3	$\frac{1}{2m\Delta}\begin{pmatrix} \Delta^2+q^2, & 2p^3p^- \\ 2p^3p^+, & -\Delta^2-q^2 \end{pmatrix}$	$\frac{1}{m}\begin{pmatrix} 0 & p^- \\ -p^+ & 0 \end{pmatrix}$	$-\frac{1}{m}\begin{pmatrix} 0 & p^- \\ -p^+ & 0 \end{pmatrix}$	$-\frac{1}{2m\Delta}\begin{pmatrix} \Delta^2+q^2, & 2p^3p^- \\ 2p^3p^+, & -(\Delta^2+q^2) \end{pmatrix}$

Nonrelativistic approximation

γ^1	$\frac{p^1}{m}\begin{pmatrix} 1 & 0 \\ 0 & 1 \end{pmatrix}$	$\begin{matrix} 0 & 1 \\ 1 & 0 \end{matrix}$	$\begin{matrix} 0 & 1 \\ 1 & 0 \end{matrix}$	$\frac{p^1}{m}\begin{pmatrix} 1 & 0 \\ 0 & 1 \end{pmatrix}$

Γ	$\tilde{u}_\alpha^{(r)} \Gamma_{\alpha\beta} u_\beta^{(s)}$	$\tilde{u}_\alpha^{(r)} \Gamma_{\alpha\beta} v_\beta^{(s)}$	$\tilde{v}_\alpha^{(r)} \Gamma_{\alpha\beta} u_\beta^{(s)}$	$\tilde{v}_\alpha^{(r)} \Gamma_{\alpha\beta} v_\beta^{(s)}$
γ^2	$\frac{p^2}{m} \begin{pmatrix} 1 & 0 \\ 0 & 1 \end{pmatrix}$	$\begin{matrix} 0 & -i \\ i & 0 \end{matrix}$	$\begin{matrix} 0 & -i \\ i & 0 \end{matrix}$	$\frac{p^2}{m} \begin{pmatrix} 1 & 0 \\ 0 & 1 \end{pmatrix}$
γ^3	$\frac{p^3}{m} \begin{pmatrix} 1 & 0 \\ 0 & 1 \end{pmatrix}$	$\begin{matrix} 1 & 0 \\ 0 & -1 \end{matrix}$	$\begin{matrix} 1 & 0 \\ 0 & -1 \end{matrix}$	$\frac{p^3}{m} \begin{pmatrix} 1 & 0 \\ 0 & 1 \end{pmatrix}$
σ^1	$\begin{matrix} 0 & 1 \\ 1 & 0 \end{matrix}$	0	0	$- \begin{pmatrix} 0 & 1 \\ 1 & 0 \end{pmatrix}$
σ^2	$\begin{matrix} 0 & -i \\ i & 0 \end{matrix}$	0	0	$- \begin{pmatrix} 0 & -i \\ i & 0 \end{pmatrix}$
σ^3	$\begin{matrix} 1 & 0 \\ 0 & -1 \end{matrix}$	0	0	$- \begin{pmatrix} 1 & 0 \\ 0 & -1 \end{pmatrix}$
$i\gamma^1\gamma^2\gamma^3$	0	$\begin{matrix} 1 & 0 \\ 0 & 1 \end{matrix}$	$\begin{matrix} 1 & 0 \\ 0 & 1 \end{matrix}$	0